LINEAR ELECTRIC ACTUATORS
AND GENERATORS

LINEAR ELECTRIC ACTUATORS AND GENERATORS

I. Boldea
Polytechnic Institute, Timisoara, Romania

Syed A. Nasar
University of Kentucky

CAMBRIDGE
UNIVERSITY PRESS

PUBLISHED BY THE PRESS SYNDICATE OF THE UNIVERSITY OF CAMBRIDGE
The Pitt Building, Trumpington Street, Cambridge, CB2 1RP, United Kingdom

CAMBRIDGE UNIVERSITY PRESS
The Edinburgh Building, Cambridge CB2 2RU, United Kingdom
40 West 20th Street, New York, NY 10011-4211, USA
10 Stamford Road, Oakleigh, Melbourne 3166, Australia

First published 1997

Printed in the United States of America

Typeset in Times

Library of Congress Cataloging-in-Publication Data
Boldea, I.
Linear electric actuators and generators / I. Boldea, Sayed A.
Nasar
p.   cm.
Includes bibliographical references and index.
ISBN 0–521–48017–5 (hc)
1. Electric generators.   2. Actuators.   I. Nasar, S. A.
II. Title.
TK2411.B6   1997
629.8'315 – dc20   96-31469
                                              CIP

*A catalog record for this book is available from
the British Library.*

ISBN 0 521 48017 5 hardback

CONTENTS

Preface   ix

CHAPTER 1
Magnetic Circuits, Fields, and Forces   1

1.1   A Review of Electromagnetic Field Theory   2
1.2   Magnetic Materials   7
1.3   Magnetic Losses   11
1.4   Magnetic Circuits   13
1.5   Ampere's Law Applied to a Magnetic Circuit   16
1.6   Limitations of the Magnetic Circuit Approach   17
1.7   Energy Relations in a Magnetic Field   19
1.8   Inductance   19
1.9   Magnetic Circuits Containing Permanent Magnets   22
1.10 Forces of Electromagnetic Origin   27
1.11 The Force Equation   28
          References   32

CHAPTER 2
Introduction to Linear Electric Actuators and Generators　　**33**

2.1	Terminology	33
2.2	Operation of LEAs	36
2.3	Operation of LEGs	39
2.4	Existing and Potential Applications	40
	References	43

CHAPTER 3
Linear Induction Actuators　　**45**

3.1	Performance Specifications	45
3.2	Construction Details for Flat and Tubular Geometries	46
3.3	Field Distributions	54
3.4	Lumped Parameter Representation	58
3.5	Steady-State Characteristics	61
3.6	State-Space Equations	68
3.7	Vector Control Aspects	70
3.8	Design Methodology by Example	75
	References	88

CHAPTER 4
Linear Permanent Magnet Synchronous Actuators　　**91**

4.1	Construction Details	91
4.2	Fields in LPMSAs	92
4.3	Sinusoidal Current Model Field Distribution	94
4.4	The *dq*-Model and Forces	97
4.5	Rectangular Current Mode	102
4.6	Dynamics and Control Aspects	105
4.7	Design Methodology	113
	References	132

CHAPTER 5
Linear Reluctance Synchronous Actuators　　**135**

5.1	Practical Configurations and Saliency Coefficients	136
5.2	Field Distributions and Saliency Ratio	141
5.3	Mathematical Model	144
5.4	Steady-State Characteristics	147
5.5	Vector Control Aspects	151
5.6	Design Methodology	152
	References	161

CHAPTER 6
Linear Switched Reluctance Actuators   **163**

6.1   Practical Configurations   163
6.2   Instantaneous Thrust   167
6.3   Average Thrust   170
6.4   Converter Rating   171
6.5   State-Space Equations   172
6.6   Control Aspects   172
6.7   Design Methodology   173
         References   178

CHAPTER 7
Linear Stepper Actuators   **179**

7.1   Practical Configurations and Their Operation   180
7.2   Static Forces   183
7.3   Lumped Parameter Models   184
7.4   Control Aspects   188
7.5   LSA Design Guidelines   193
         References   199

CHAPTER 8
Linear Electric Generators   **201**

8.1   Principle of Operation and Basic Configurations   201
8.2   Moving Magnet Linear Alternators   203
8.3   Moving Iron Linear Alternators   210
8.4   Stability Considerations   223
8.5   Choice of Tuning Capacitor   232
         References   233

Index   **235**

PREFACE

Linear electric actuators are electromagnetic devices capable of producing directly (without any linkages, etc.) progressive unidirectional or oscillatory short-stroke motion. The motion occurs because of the electromagnetic force developed in the actuator. Linear electric generators are also linear motion electromagnetic devices which transform short-stroke oscillatory motion mechanical energy into single-phase ac electrical energy. Just as a rotary electric machine may operate either as a motor or as a generator, a linear motion electromagnetic device may be designed to work either as an actuator or as a generator. From this standpoint, linear electric actuators and generators are the counterparts of a corresponding rotary electric machine. In general, however, linear electric machines have been associated with long linear progressive motion, such as in transportation and similar applications.

Whereas primitive linear electric machines have been in existence for a long time, since the 1960s there has been a great deal of interest in linear machines for various applications—especially transportation. Several books and numerous papers have been published on the subject in the recent past. However, the literature on linear electric actuators and generators is relatively sparse. Clearly, the potential applications of these devices are too numerous to mention here. Judged from the present trend, it would suffice to say that the field of linear actuators and generators may lead to an industry of "linear motion control" with a large worldwide market.

Much of the existing literature on the subject of linear actuators and generators deals with the principles of operation and performance calculations. Not much has been published on their control and detailed design methodologies. In this book, we present a unified treatment combining topologies of these devices with pertinent field distributions obtained by the finite-element method, state-space equations governing their dynamics and control, and detailed design methodologies. The book contains much new and original material developed by the authors.

In Chapter 1, we briefly review magnetic circuits, fields, and forces. In Chapter 2, we present an overview of linear electric actuators and generators (LEAGs). Basic definitions, terminology, and operating principles of LEAGs are introduced in this chapter. A classification of these devices is also presented in Chapter 2. Linear induction actuators are discussed in Chapter 3. Chapters 4 and 5, respectively, present linear permanent magnet and linear reluctance synchronous actuators. Linear switched reluctance actuators are described in Chapter 6, linear stepper actuators in Chapter 7, and linear electric generators (also termed linear alternators) in Chapter 8. We have not included "solenoids" here, as they are widely discussed in the literature. Throughout the book we have attempted to give a unified presentation, namely, basic construction and topology, field distribution, lumped parameter equivalent circuit, state-space equations, dynamics and control, followed by design examples.

CHAPTER

1

MAGNETIC CIRCUITS, FIELDS, AND FORCES

In this book we deal with magnetic and electromagnetic devices. Specifically we discuss the theory, design, and analysis of linear motion electric actuators and generators. Whereas actuators are not precisely defined, we may consider them to be as devices which convert a form of energy into controlled mechanical motion. The form of energy may be electric, hydraulic, or pneumatic. In this book we are concerned with electric actuators, which convert electric energy into controlled mechanical motion of limited travel. Electric actuators compare favorably with hydraulic and pneumatic actuators, and are much superior to both in terms of efficiency, controllability, cost, and environmental safety. Of course, electric actuators may be either rotary generating rotary motion or linear leading to linear motion. Furthermore, a linear electric actuator may be direct operating, utilizing electromagnetic or piezoelectric effects to produce force and motion. On the other hand, a converting type of linear actuator uses an electric motor with gears and linkages to produce linear motion. The direct operating type of linear actuators are the subject matter of this book. Most of these devices invariably consist of coupled magnetic and electric circuits in relative motion, although piezoelectric actuators operate on a different principle.

Because the electromechanical energy conversion process is reversible, actuators may be operated as linear electric generators, in which case mechanical energy is transformed into electric energy. The fundamental energy conversion equation is [1]

$$F\dot{x} = vi \qquad\qquad (1.1)$$

where F is mechanical force, N; $\dot{x}$ is mechanical velocity, m/s; v is voltage, V; and i is current, A. In (1.1) it is assumed that F and $\dot{x}$ are in the same direction. Again, like actuators, linear electric generators are most often electromagnetic linear motion devices.

So in this chapter we briefly review the laws governing the magnetic circuit and magnetic fields. Because actuators and generators have inherent forces of magnetic and/or electromagnetic origin, we also present various methods of determining these forces. Subsequently we consider examples that illustrate the applications of the procedure developed in this chapter.

1.1 A REVIEW OF ELECTROMAGNETIC FIELD THEORY

The basic laws of electricity are governed by a set of equations called *Maxwell's equations*. Naturally, these equations govern the electromagnetic phenomena in energy conversion devices too. Because of the presence of moving media, a direct application of Maxwell's equations to energy conversion devices is rather subtle. But certain other aspects of applications, such as parameter determination, are straight forward. In this section, we briefly review Maxwell's equations, then derive certain other equations from them, and finally discuss some examples of applications of Maxwell's equations and of other equations derived therefrom.

We know that charged particles are acted upon by forces when placed in *electric* and *magnetic fields*. In particular, the magnitude and direction of the force $\mathbf{F}$ acting on a charge q moving with a velocity $\mathbf{u}$ in an electric field $\mathbf{E}$ and in a magnetic field $\mathbf{B}$ is given by the Lorentz force equation:

$$\mathbf{F} = q(\mathbf{E} + \mathbf{u} \times \mathbf{B}) = \mathbf{F}_E + \mathbf{F}_B \qquad\qquad (1.2)$$

The electric field intensity $\mathbf{E}$ is thus defined by the following equation:

$$\mathbf{E} = \frac{\mathbf{F}_E}{\Delta q} \qquad\qquad (1.3)$$

where $\mathbf{F}_E$ is the force on an infinitesimal test charge Δq; the electric field is measured in volts per meter. The magnetic field can also be similarly defined, from (1.2), as the force on a unit charge moving with unit velocity at right angles to the direction of $\mathbf{B}$. The quantity $\mathbf{B}$ is called the *magnetic flux density* and is measured in webers per square meter, or tesla (T).

Having defined the electric field, we can now define the *potential difference dV* between two points separated by a distance $d\mathbf{l}$ as

$$dV = -\mathbf{E} \cdot d\mathbf{l} \tag{1.4}$$

Or, using the vector operator ∇, (1.4) is expressed as

$$\mathbf{E} = -\nabla V \tag{1.5}$$

We also see that, if **B** is the magnetic flux density, the total *magnetic flux* ϕ can be expressed as

$$\phi = \int_s \mathbf{B} \cdot d\mathbf{s} \tag{1.6}$$

where the integral is over a surface s.

With these definitions in mind, we now recall Faraday's law. It states that an electromotive force (emf) is induced in a closed circuit when the magnetic flux ϕ linking the circuit changes. If the closed circuit consists of an N-turn coil, the induced emf is given by

$$\mathrm{emf} = -N\frac{d\phi}{dt} \tag{1.7}$$

The negative sign is introduced to take into account Lenz's law and to be consistent with the positive sense of circulation about a path with respect to the positive direction of flow through the surface (as shown in Fig. 1.1) when (1.7) is expressed in the integral form as follows.

Because the potential difference has been defined by (1.4), it is reasonable to define emf as

$$\mathrm{emf} = \oint \mathbf{E} \cdot d\mathbf{l} \tag{1.8}$$

which can be considered to be the potential difference about a specific closed path. If we consider "going around the closed path only once," we can put $N = 1$; then from (1.6) to (1.8) it follows that

$$\oint \mathbf{E} \cdot d\mathbf{l} = -\frac{\partial}{\partial t} \int_s \mathbf{B} \cdot d\mathbf{s} \tag{1.9}$$

The partial derivative with respect to time is used to distinguish derivatives with space, since **B** can be a function of both time and space.

Equation (1.9) is an expression of Faraday's law in the integral form. We now recall *Stokes' theorem*, which states that

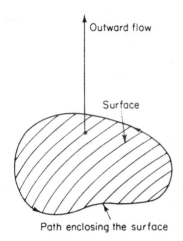

Outward flow

Surface

Path enclosing the surface

FIGURE 1.1
Positive circulation about a path enclosing a surface and positive flow through the surface.

$$\oint \mathbf{E} \cdot d\mathbf{l} = \int_s (\nabla \times \mathbf{E}) \cdot d\mathbf{s} \tag{1.10}$$

where the surface **s** is that enclosed by the closed path, as shown in Fig. 1.1. We notice that the terms on the left-hand sides of (1.9) and (1.10) are identical. Equating the terms on the right-hand sides of these equations we have

$$\nabla \times \mathbf{E} = -\frac{\partial \mathbf{B}}{\partial t} \tag{1.11}$$

which is an expression of Faraday's law in the differential form.

A knowledge of vector algebra shows that the divergence of a curl is zero. Expressed mathematically,

$$\nabla \cdot \nabla \times \mathbf{E} = 0 \tag{1.12}$$

From (1.11) and (1.12), therefore,

$$\nabla \cdot \mathbf{B} = 0 \tag{1.13}$$

In deriving (1.11) we assumed that there was no relative motion between the closed circuit and the magnetic field. If, however, there is relative motion between the circuit and the magnetic field, in addition to the time variation of this field, (1.11) has to be modified. Referring to the Lorentz force equation, (1.2), we notice that the fields **E** and **B** are measured in a stationary reference frame. It is only the charge that is moving with a velocity **u**. If, on the other hand, the charge moves with a velocity **u'** with respect to a moving reference frame having a velocity **u**, the velocity of the charge with respect to the stationary reference frame becomes (**u** + **u'**). The force on the charge is then given by

$$\mathbf{F} = q[\mathbf{E} + (\mathbf{u} + \mathbf{u}') \times \mathbf{B}] = q[(\mathbf{E} + \mathbf{u} \times \mathbf{B}) + \mathbf{u}' \times \mathbf{B}] \qquad (1.14)$$

Thus, the electric field **E'** in the moving reference frame becomes

$$\mathbf{E}' = \mathbf{E} + \mathbf{u} \times \mathbf{B} \qquad (1.15)$$

where **E** is given by (1.11). Here we have not taken into account relativistic effects because in electromagnetic energy conversion devices the velocities involved are considerably small. From (1.11) and (1.15) we see that the electric field **E** induced in a circuit moving with a velocity **u** in a time-varying field **B** is given by

$$\nabla \times \mathbf{E} = -\frac{\partial \mathbf{B}}{\partial t} + \nabla \times \mathbf{u} \times \mathbf{B} \qquad (1.16)$$

In (1.16) we notice that the second term on the right-hand side is introduced to take into account the motion of the circuit. Thus, we can identify the first term in (1.16) as due to a *transformer emf* and the second as due to a *motional emf;* that is, the "change in the flux linkage" rule and the "flux-cutting" rule of obtaining the emf induced in a circuit are both taken into account in (1.16).

We would like to point out here that (1.7), expressing Faraday's law, is complete. A slightly more tedious analysis than that given above can be shown to lead to (1.16). In practical cases, it is better to identify the transformer and motional voltages separately. A direct application of (1.7), without extra care, may lead to inconsistent results.

So far we have discussed ways of finding the electric field from given magnetic fields. We shall now consider the relationship between given currents and resulting magnetic fields. In this connection, *Ampere's circuital law* expressed as

$$\oint \mathbf{H} \cdot d\mathbf{l} = I \qquad (1.17)$$

gives the relationship between the *magnetic field intensity* **H** and the current I enclosed by the closed path of integration. If **J** is the surface current density, (1.17) can be also written as

$$\oint \mathbf{H} \cdot d\mathbf{l} = \int_s \mathbf{J} \cdot d\mathbf{s} \tag{1.18}$$

Using Stokes' theorem, (1.10), the point form of Ampere's law becomes, from (1.18),

$$\nabla \times \mathbf{H} = \mathbf{J} \tag{1.19}$$

The surface current density $\mathbf{J}$ is related to the volume charge density ρ through the *continuity equation*:

$$\nabla \cdot \mathbf{J} = -\frac{\partial \rho}{\partial t} \tag{1.20}$$

Taking the divergence of (1.19) reveals an immediate inconsistency when compared with (1.20), since $\nabla \cdot \nabla \times \mathbf{H} = 0$. If an extra term $\partial \mathbf{D}/\partial t$ is added to the right-hand side of (1.19) such that

$$\nabla \times \mathbf{H} = \mathbf{J} + \frac{\partial \mathbf{D}}{\partial t} \tag{1.21}$$

we find that the continuity equation is satisfied, provided that

$$\nabla \cdot \mathbf{D} = \rho \tag{1.22}$$

But (1.22) is perfectly legitimate, since it expresses *Gauss' law* in the differential form. The quantity $\mathbf{D}$ is called the *electric flux density* and $\partial \mathbf{D}/\partial t$ is known as the *displacement current* density.

From the preceding considerations we have the following set of equations relating to the various field quantities:

$$\nabla \times \mathbf{E} = -\frac{\partial \mathbf{B}}{\partial t} \tag{1.23}$$

$$\nabla \times \mathbf{H} = \mathbf{J} + \frac{\partial \mathbf{D}}{\partial t} \tag{1.24}$$

$$\nabla \cdot \mathbf{B} = 0 \tag{1.25}$$

$$\nabla \cdot \mathbf{D} = \rho \tag{1.26}$$

Equations (1.23) through (1.26) are generally known as *Maxwell's equations.*

In order to obtain complete information regarding the various field quantities, in addition to Maxwell's equations certain auxiliary relations are also very useful. These relations are as follows.

Ohm's law. For a conductor of conductivity σ,

$$\mathbf{J} = \sigma\mathbf{E} \qquad (1.27)$$

where $\mathbf{J}$ is surface current density and $\mathbf{E}$ electric field intensity.

Permittivity. The electric field intensity and the electric flux density in a medium are related to each other by

$$\mathbf{D} = \varepsilon\mathbf{E} \qquad (1.28)$$

where ε is called the *permittivity* of the material.

Permeability. The magnetic field intensity and the magnetic flux density in a material are related to each other by

$$\mathbf{B} = \mu\mathbf{H} \qquad (1.29)$$

where μ is called the *permeability* of the medium.

Finally, we should note that in the majority of energy conversion devices there are no free charges. Consequently, for these cases the Lorentz force equation takes the form

$$\mathbf{F} = \mathbf{J} \times \mathbf{B} \qquad (1.30)$$

which follows from (1.2), since the motion of charges constitutes the flow of current. Because $\mathbf{J}$ and $\mathbf{B}$ are the current and flux densities, respectively, (1.30) determines the force density rather than the total force.

Because most commonly encountered electric actuators have magnetic structures, we now consider some properties of magnetic materials.

1.2 MAGNETIC MATERIALS

Returning to (1.29), in free space $\mathbf{B}$ and $\mathbf{H}$ are related by the constant μ_0, known as the *permeability of free space*:

$$\mathbf{B} = \mu_0 \mathbf{H} \tag{1.31}$$

and

$$\mu_0 = 4\pi \times 10^{-7} \text{ H/m}$$

Within a material, (1.31) must be modified to describe a magnetic phenomenon different from that occurring in free space:

$$\mathbf{B} = \mu\mathbf{H}, \qquad \mu = \mu_R\mu_0 \tag{1.32}$$

where μ is termed *permeability* and μ_R *relative permeability*, a nondimensional constant. Permeability in a material medium must be further qualified as applicable only in regions of homogeneous (uniform quality) and *isotropic* (having the same properties in any direction) materials. In materials not having these characteristics, μ becomes a vector. Finally, note that for some common materials, (1.29) is *nonlinear*, and μ varies with the magnitude of **B**. This results in several subdefinitions of permeability related to the nonlinear *B-H* characteristic of the material, which will be discussed next.

A material is classified according to the nature of its relative permeability μ_R, which is actually related to the internal atomic structure of the material and will not be discussed further at this point. Most "nonmagnetic" materials are classified as either *paramagnetic*, for which μ_R is slightly greater than 1.0, or *diamagnetic*, for which μ_R is slightly less than 1.0. However, for all practical purposes, μ_R can be considered equal to 1.0 for all of these materials.

Because of the nonlinear characteristics of most magnetic materials, graphical techniques are generally valuable in describing their magnetic characteristics. The two graphical characteristics of most importance are known as the *B-H* curve, or magnetization characteristic, and the hysteresis loop. Figure 1.2 shows a typical *B-H* characteristic. This characteristic can be obtained in two ways: the *virgin B-H* curve, obtained from a totally demagnetized sample, or the *normal B-H* curve, obtained as the tips of hysteresis loops of increasing magnitude. There are slight differences between the two methods that are not important for our purposes. The *B-H* curve is the result of *domain* changes within the magnetic material. Ferromagnetic materials are divided into small regions or domains (approximately 10^{-2} to 10^{-5} cm in size); in each region all dipole moments are spontaneously aligned. When the material is completely demagnetized, these domains have random orientation resulting in zero net flux density in any finite sample. As an external magnetizing force, *H*, is applied to the material, the domains that happen to be in line with the direction of *H* tend to grow, increasing *B* (region I in Fig. 1.2). In region II, as *H* is further increased, the domain walls move rapidly until each crystal of the material is a single domain. In region III the domains rotate in some direction until all domains are aligned with *H*. This results in *magnetic saturation*, and the flux density within the material cannot increase beyond the *saturation density* B_s. The small increase that

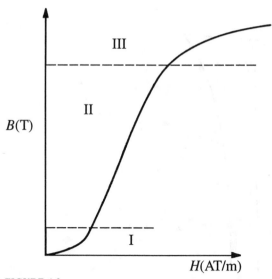

FIGURE 1.2
A typical *B-H* curve.

occurs beyond this condition is due to the increase in the space occupied by the material according to the relationship $B = \mu_0 H$. It is often convenient to subtract out this component of "free space" flux density and observe only the flux density variation within the material. Such a curve is known as the *intrinsic magnetization curve* and is of use in the design of permanent magnet devices.

The regions shown in Fig. 1.2 are also of value in describing the nonlinear permeability characteristic. From (1.32) it is seen that permeability is the slope of the *B-H* curve. In the following discussion relative permeability is assumed; that is, the factor μ_0 is factored out. The slope of the *B-H* curve is actually properly called *relative differential permeability*, or

$$\mu_d = \frac{1}{\mu_0} \frac{dB}{dH} \qquad (1.33)$$

Relative initial permeability is defined as

$$\mu_i = \lim_{H \to 0} \frac{1}{\mu_0} \left| \frac{B}{H} \right| \qquad (1.34)$$

and is seen to be the permeability in region I. In region II the *B-H* curve for many materials is relatively straight, and if a magnetic device is operated only in this region, linear theory can be used. In all regions the most general permeability term is known

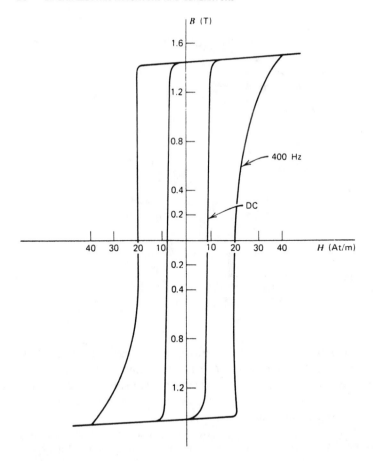

FIGURE 1.3
Deltamax tape-wound core 0.002-in. strip hysteresis loop.

as *relative amplitude permeability* and is defined as merely the ratio of B to H at any point on the curve, or

$$\mu_a = \frac{1}{\mu_0} \left| \frac{B}{H} \right| \qquad (1.35)$$

In general, permeability has to be defined on the basis of the type of signal exciting the magnetic material.

The second graphical characteristic of interest is the hysteresis loop; a typical sample is shown in Fig. 1.3. This is a *symmetrical hysteresis loop*, obtained only after a number of reversals of the magnetizing force between plus and minus H_s. This characteristic illustrates several parameters of most magnetic materials, the most obvious being the property of hysteresis itself. The area within the loop is related to

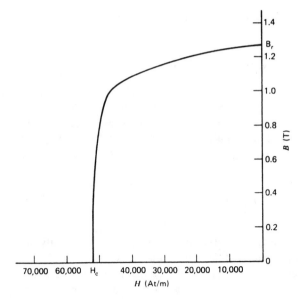

FIGURE 1.4
Demagnetization curve of Alnico V.

the energy required to reverse the magnetic domain walls as the magnetizing force is reversed. This is a nonreversible energy and results in an energy loss known as the *hysteresis loss.* This area varies with temperature and the frequency of reversal of H in a given material.

The second quadrant of the hysteresis loop, shown in Fig. 1.4 for Alnico V magnet, is very valuable in the analysis of devices containing permanent magnets. The intersection of the loop with the hori-zontal (H) axis is known as the *coercive force* H_c and is a measure of the magnet's capacity to withstand demagnetization from external magnetic signals. Often shown on this curve is a second curve, known as the *energy product,* which is the product of B and H plotted as a function of H and is a measure of the energy stored in the permanent magnet. The value of B at the vertical axis is known as the *residual flux density.*

The *Curie temperature,* or Curie point, T_c, is the critical temperature above which a ferromagnetic material becomes paramagnetic.

1.3 MAGNETIC LOSSES

A characteristic of magnetic materials that is very significant in the energy efficiency of an electromagnetic device is the energy loss within the magnetic material itself. In electric machines and transformers, this loss is generally termed the *core loss* or sometimes the *magnetizing loss* or *excitation loss.* Traditionally, core loss has been divided into two components: *hysteresis loss* and *eddy current loss.* The hysteresis loss component has been alluded to previously and is generally held to be equal to the area of the low-frequency hysteresis loop times the frequency of the magnetizing force in sinusoidal systems. The hysteresis loss P_h is given by the empirical relationship

$$P_h = k_h f B_m^{1.5 \text{ to } 2.5} \text{ W/kg} \tag{1.36}$$

where k_h is a constant, f the frequency, and B_m the maximum flux density.

Eddy current losses are caused by induced electric currents, called eddies, since they tend to flow in closed paths within the magnetic material itself. The eddy current loss in a sinusoidally excited material, neglecting saturation, can be expressed by the relationship

$$P_e = k_e f^2 B_m^2 \text{ W/kg} \tag{1.37}$$

where B_m is the maximum flux density, f the frequency, and k_e a proportionality constant, depending on the type of material and the lamination thickness.

To reduce the eddy current loss, the magnetic material is *laminated*, that is, divided into thin sheets with a very thin layer of electrical insulation between the sheets. The sheets must be oriented in a direction parallel to the flow of magnetic flux (Fig. 1.5). The eddy current loss is roughly proportional to the square of the lamination thickness and inversely proportional to the electrical resistivity of the material. Lamination thickness varies from about 0.5 to 5 mm in electromagnetic devices used in power applications and from about 0.01 to 0.5 mm in devices used for electronics applications. Laminating a magnetic part usually increases its volume. This increase may be appreciable, depending on the method used to bond the laminations together. The ratio of the volume actually occupied by magnetic material to total volume of a magnetic part is the *stacking factor*. This factor is important in accurately calculating flux densities in magnetic parts. Table 1.1 gives typical stacking factors for the thinner lamination sizes. Stacking factor approaches 1.0 as the lamination thickness increases.

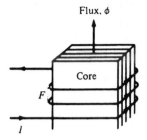

FIGURE 1.5
A portion of a laminated core.

TABLE 1.1
Stacking Factor for Laminated Cores

Lamination Thickness (mm)	Stacking Factor
0.0127	0.50
0.0254	0.75
0.0508	0.85
0.1-0.25	0.90
0.27-0.36	0.95

1.4 MAGNETIC CIRCUITS

It is important to emphasize that a magnetic field is a distributed parameter phenomenon; that is, it is distributed over a region of space. As such, rigorous analysis requires the use of the distance variables as contained in the divergence and curl symbols of (1.11) and (1.13), for instance. However, under the proper conditions, it is possible to apply lumped parameter analysis to certain classes of magnetic field problems just as it is applied in electric circuit analysis.

This section briefly describes lumped circuit analysis as applied to magnetic systems, often called *magnetic circuit analysis*. Magnetic circuit analysis follows the approach of simple dc electric circuit analysis and applies to systems excited by dc signals or, by means of an incremental approach, to low-frequency ac excitation. Its usefulness lies in sizing the magnetic components of an electromagnetic device during design stages, calculating inductances, and determining airgap flux density for power and force calculations.

Let us begin with a few definitions.

1. *Magnetic Potential.* For regions in which no electric current densities exist, which is true for the magnetic circuits we will discuss, the magnetic field intensity **H** can be defined in terms of *scalar* magnetic potential F as

$$\mathbf{H} = \nabla F; \quad F = \int \mathbf{H} \cdot d\mathbf{l} \tag{1.38}$$

It is seen that F has the dimension of amperes, although "ampere-turns" is frequently used as a unit for F. For a potential rise or source of magnetic energy, the term *magnetomotive force* (mmf) is frequently used. As a potential drop, the term *reluctance drop* is often used. There are two types of sources of mmf in magnetic circuits: electric current and permanent magnets. The current source usually consists of a coil of a number of turns, N, carrying a current known as the *exciting current*. Note that N is nondimensional.

2. *Magnetic Flux.* Streamlines or flowlines in a magnetic field are known as lines of magnetic flux, denoted by the symbol ϕ, and having the SI unit weber. Flux is related to **B** by the surface integral

$$\phi = \int_s \mathbf{B} \cdot d\mathbf{s} \tag{1.39}$$

3. *Reluctance.* Reluctance is a component of magnetic impedance, somewhat analogous to resistance in electric circuits except that reluctance is not an energy loss component. It is defined by a relationship analogous to Ohm's law:

$$\phi = \frac{F}{\mathfrak{R}} \tag{1.40}$$

The SI unit of magnetic reluctance is henry^{-1}. In regions containing magnetic material that is homogeneous and isotropic and where the magnetic field is uniform, (1.40) gives further insight into the nature of reluctance. If we assume that the flux density has only one directional component, B, and is uniform over a cross section of area A_m, taken perpendicular to the direction of B, (1.39) becomes $\phi = BA_m$. We also assume that H is nonvarying along the length l_m in the direction of B, and (1.40) becomes, with some rearranging,

$$\mathfrak{R} = \frac{F}{\phi} = \frac{Hl_m}{BA_m} = \frac{l_m}{\mu A_m} \tag{1.41}$$

which is similar to the expression for electrical resistance in a region with similarly uniform electrical properties.

4. *Permeance.* The permeance $\wp$ is the reciprocal of reluctance and has the SI unit henry. Permeance and reluctance are both used to describe the geometric characteristics of a magnetic field, mainly for purposes of calculating inductances.

5. *Leakage Flux.* Between any two points at different magnetic potentials in space, a magnetic field exists, as shown by (1.38). In any practical magnetic circuit there are many points or, more generally, planes at magnetic potentials different from each other. The magnetic field between these points can be represented by flowlines or lines of magnetic flux. Where these flux lines pass through regions of space—generally air space, electrical insulation, or structural members of the system—instead of along the main path of the circuit, they are termed *leakage flux*

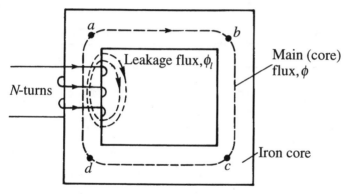

FIGURE 1.6
Leakage flux.

lines. In coupled circuits with two or more windings the definition of leakage flux is specific: flux links one coil but not the other. Leakage flux is identified, in Fig. 1.6, along the path l whereas the main core flux, also termed mutual flux, is along the path $abcd$.

Leakage is a characteristic of all magnetic circuits and can never be completely eliminated. At dc or very low ac excitation frequencies, magnetic shielding consisting of thin sheets of high-permeability material can reduce leakage flux. This is done not by eliminating leakage but by establishing new levels of magnetic potential in the leakage paths to better direct the flux lines along the desired path. At higher frequencies of excitation, electrical shielding, such as aluminum foil, can reduce leakage flux by dissipating its energy as induced currents in the shield.

6. *Fringing*. Fringing is somewhat similar to leakage; it describes the spreading of flux lines in an airgap of a magnetic circuit. Figure 1.7 illustrates fringing at a gap. Fringing results from lines of flux that appear along the sides and edges of the two

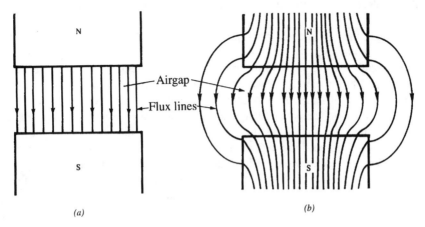

FIGURE 1.7
(a) No fringing of flux; *(b)* fringing of flux.

TABLE 1-2
Analogy between a Magnetic Circuit and a dc Electric Circuit

Magnetic Circuit	Electric Circuit
Flux, ϕ	Current, I
mmf, F	Voltage, V
Reluctance, $\Re = l/\mu A$	Resistance, $R = l/\sigma A$
Permeance, $\wp = 1/\Re$	Conductance, $G = 1/R$
Permeability, μ	Conductivity, σ
Ohm's law, $\phi = F/\Re$	Ohm's law, $I = V/R$

magnetic members at each side of the gap, which are at different magnetic potentials. Fringing is almost impossible to calculate analytically, except in the simplest of configurations. Fringing has the effect of increasing the effective area of the airgap, which must be considered with the length of the airgap.

Based on (1.40) and (1.41), we may now construct the relationships summarized in Table 1.2. In this table l is the length and A is the cross-sectional area of the path for the flow of current in the electric circuit or for the flux in the magnetic circuit. It may be verified from Table 1.2 that the unit of reluctance is henry^{-1}.

Because ϕ is analogous to I and $\Re$ is analogous to R, the laws for series- or parallel-connected resistors also hold for reluctances.

1.5 AMPERE'S LAW APPLIED TO A MAGNETIC CIRCUIT

According to (1.17), the integral around any closed path of the magnetic field intensity **H** equals the electric current contained within that path. A word about directions in using this integral expression is in order here. Positive current is defined as flowing in the direction of the advance of a right-handed screw turned in the direction in which the closed path is traversed.

Let us apply Ampere's law to the simple magnetic circuit whose cross section is shown in Fig. 1.8; the circuit consists of a magnetic member of mean length l_m, in series with an airgap of length l_g, around which are wrapped three coils of turns N_1, N_2, and N_3 respectively. The path of magnetic flux ϕ is shown along the mean length of the magnetic member and across the airgap. Let the line integration proceed in a clockwise manner. Current directions are shown in the three coils. Note that for the directions shown, current direction is into the plane of the paper for the conductors enclosed by the integration paths for coils 1 and 3 and out of the plane of the paper for coil 2. From (1.17) we obtain (after neglecting saturation and leakage)

$$\oint \mathbf{H} \cdot d\mathbf{l} = H_m l_m + H_g l_g = \phi(\Re_m + \Re_g) = F_m + F_g = I \qquad (1.42)$$

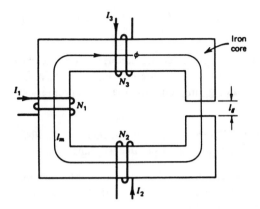

FIGURE 1.8
A composite magnetic circuit, with multiple excitation (mmf's).

where $\mathfrak{R}_m$ and $\mathfrak{R}_g$ are the reluctances of the magnetic member and gap, respectively, and F_m and F_g represent the magnetic potential or reluctance drop across these two members of the magnetic circuit. The right side of (1.17) gives

$$I = N_1 I_1 + N_3 I_3 - N_2 I_2 \tag{1.43}$$

Combining (1.41) and (1.42) yields

$$N_1 I_1 + N_3 I_3 - N_2 I_2 - F_m - F_g = 0 \tag{1.44}$$

We may generalize Ampere's law on the basis of this simple example to state that the sum of the magnetic potentials around any closed path is equal to zero, which is analogous to Kirchhoff's voltage relationship in electric circuits.

1.6 LIMITATIONS OF THE MAGNETIC CIRCUIT APPROACH

The number of problems in practical magnetic circuits that can be solved by the approach outlined in Sections 1.4 and 1.5 is limited, despite the similarity of this approach to simple dc electric circuit theory. The purpose of introducing magnetic circuits is to state some very fundamental principles and definitions that are necessary in order to understand electromagnetic systems. The limitations of magnetic circuit theory rest primarily in the nature of magnetic materials as contrasted with conductors, insulators, and dielectric materials. Most of these limitations have already been introduced as "assumptions" in the discussion of magnetic circuits. Let us assess the significance of these assumptions.

1. *Homogeneous Magnetic Material.* Most materials used in practical electromagnetic systems can be considered homogeneous over finite regions of space, allowing the

use of the integral forms of Maxwell's equations and calculations of reluctances and permeances.

2. *Isotropic Magnetic Materials.* Many sheet steels and ferrites are oriented by means of the metallurgical process during their production. Oriented materials have a "favored" direction in their grain structure, giving superior magnetic properties when magnetized along this direction.

3. *Nonlinearity.* This is an inherent property of all ferro- and ferrimagnetic materials. However, there are many ways of treating this class of nonlinearity analytically.

   (a) As can be seen from the *B-H* curve shown in Fig. 1.2, a considerable portion of the curve for most materials can be approximated as a straight line, and many electromagnetic devices operate in this region.

   (b) Numerous analytical and numerical techniques have been developed for describing the *B-H* and other nonlinear magnetic characteristics.

   (c) The nonlinear *B-H* characteristic of magnetic materials manifests itself in their relationship between flux and exciting current in electromagnetic systems; the relationship between flux and induced voltage is a *linear* one, as given by Faraday's law, (1.9). It is possible to treat these nonlinear excitation characteristics separately in many systems, such as is done in the equivalent circuit approach to transformers and induction motors.

   (d) An inductance whose magnetic circuit is composed of a magnetic material is a nonlinear electric circuit element, such as a coil wound on a magnetic toroid. With an airgap in the magnetic toroid, however, the effect of the nonlinear magnetic material on the inductance is lessened. Actuators have airgaps in their magnetic circuits, permitting the basic theory of these devices to be described by means of linear equations.

4. *Saturation.* All engineering materials and devices exhibit a type of saturation when output fails to increase with input, for instance, in the saturation of an electronic amplifier. Saturation is very useful in many electromagnetic devices, such as magnetic amplifiers and saturable reactors.

5. *Leakage and Fringing Flux.* This is a property of all magnetic circuits. It is best treated as a part of the generalized solution of magnetic field distribution in space, often called a *boundary value problem.* In many actuator magnetic circuits, boundaries between regions of space containing different types of magnetic materials (usually a boundary between a ferromagnetic material and air) are often planes or cylindrical surfaces that, in a two-dimensional cross section, become straight lines or circles. Leakage inductances can frequently be determined in such regions by calculating the reluctance or permeance of the region using fairly simple integral formulations. The spatial or geometric coefficients so obtained are known as *permeance coefficients.*

1.7 ENERGY RELATIONS IN A MAGNETIC FIELD

The energy stored in a magnetic field is defined throughout space by the volume integral

$$W = \frac{1}{2} \int_{vol} \mathbf{B} \cdot \mathbf{H} \, dv = \frac{1}{2} \int_{vol} \mu H^2 \, dv = \frac{1}{2} \int_{vol} \frac{B^2}{\mu} \, dv \qquad (1.45)$$

This equation is valid only in regions of constant permeability. Therefore, its usefulness is limited to *static* linear magnetic circuits. Energy relationships in time-varying magnetic circuits, such as electric actuators, including the concept of *coenergy*, will be developed later in the book. (See Section 1.11).

1.8 INDUCTANCE

Inductance is one of the three circuit constants in electric circuit theory and is defined as flux linkage per ampere:

$$L = \frac{\lambda}{i} = \frac{N\phi}{i} \qquad (1.46)$$

Consider the magnetic toroid around which are wound n distinct coils electrically isolated from each other, as shown in Fig. 1.9. The coils are linked magnetically by the flux ϕ, some portion of which links each of the coils. A number of inductances can be defined for this system:

$$L_{km} = \frac{\text{flux linking the } k\text{th coil due to the current in the } m\text{th coil}}{\text{current in the } m\text{th coil}}$$

Mathematically, this can be stated as

$$L_{km} = \frac{N_k(K\phi_m)}{i_m} \qquad (1.47)$$

where K is the portion of the flux due to coil m that links coil k and is known as the *coupling coefficient*. By definition, its maximum value is 1.0. A value of K less than 1.0 is attributable to leakage flux in the regions between the location of coil k and coil m. When the two subscripts in (1.47) are identical, the inductance is termed *self-inductance*; when different, the inductance is termed *mutual inductance* between coils k and m. Mutual inductances are symmetrical; that is,

$$L_{km} = L_{mk} \qquad (1.48)$$

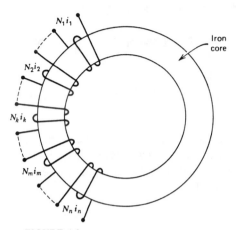

FIGURE 1.9
A toroid with n windings.

Inductance can be related to the magnetic parameters derived earlier in this chapter. In (1.47), ϕ_m can be replaced, using (1.40), by the magnetic potential of coil m, F_m, divided by reluctance of the magnetic circuit, $\Re$; the magnetic potential of coil m, however, is $N_m I_m$. Making these substitutions in (1.47) gives

$$L_{km} = \frac{K N_k N_m}{\Re} = K N_k N_m \wp \tag{1.49}$$

where $\wp$ is the permeance, the reciprocal of the reluctance.

Stored energy can be expressed in terms of inductance:

$$W = \frac{1}{2} L i^2 \tag{1.50}$$

By substituting for L from (1.46) and for Ni (magnetic potential) from (1.40), (1.50) can be expressed as

$$W = \frac{1}{2} \Re \phi^2 \tag{1.51}$$

At this point we observe that the determination of permeances of various magnetic circuit configurations plays a key role in obtaining the fluxes, inductances, etc., pertaining to the magnetic circuit. Reference [2] gives the procedures for determining permeance functions of numerous practical electromagnetic actuators. The following example calculates the slot inductance and illustrates the use of permeance coefficients in determining the slot inductance.

Example 1.1 Determine the inductance of the conductor, of length l_a into the paper, in a slot having the dimensions shown in Fig. 1.10, assuming the magnetic material to be infinitely permeable.

First we introduce the concept of *partial flux linkage*. This term is used to describe flux lines that link only a portion of an electrical conductor or only a portion of the turns of a coil. In this example, a flux line across the slot at any height below y_2 links only the current below the flux line. (The path of the flux line is closed through the magnetic material, as shown.) The assumption of horizontal flux lines is not exactly correct in a practical configuration, but an analytical solution would be impossible without this assumption.

Consider the differential flux in the "strip" at a distance y from the slot bottom; the strip has height dy, width l_a (into the paper), and length t_2. The magnetic potential enclosed by this strip is

$$F_y = Jt_2 y$$

The permeance of the strip is

$$d\mathcal{P}_y = \frac{\mu_0 l_a \, dy}{t_2}$$

The flux through the strip is, from (1.40),

$$d\phi = F_y \, d\mathcal{P}_y = \mu_0 J l_a y \, dy$$

The total flux across the gap is

$$\phi_2 = \mu_0 J l_a \int_0^{y_2} y \, dy = \mu_0 J l_a \frac{y_2^2}{2}$$

The total magnetic potential of the slot is

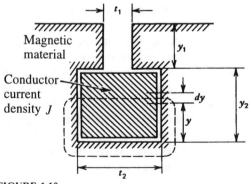

FIGURE 1.10
A slot cross section.

$$F_s = Jy_2 t_2$$

The permeance of the lower portion of the slot is

$$\wp_2 = \frac{\phi_2}{F_s} = \frac{\mu_0 l_a y_2}{2 t_2}$$

The permeance of the upper portion of the slot is

$$\wp_1 = \mu_0 y_1 \frac{l_a}{t_1}$$

The slot inductance is

$$L_s = \wp_1 + \wp_2 = \mu_0 \left(\frac{y_1 l_a}{t_1} + \frac{y_2 l_a}{2 t_2} \right) \text{H}$$

1.9 MAGNETIC CIRCUITS CONTAINING PERMANENT MAGNETS

The second type of excitation source commonly used for supplying energy to magnetic circuits used in electromechanical devices is the permanent magnet. There is obviously a great difference in physical appearance between an electrical exciting coil and a permanent magnet source of excitation, so we should expect some differences in methods of analysis used in the two types of magnetic circuits. Actually, these differences are relatively minor and are related to the use of the permanent magnet itself, not to the other portions of the magnetic circuit.

An electrical excitation coil energized from a constant-voltage or constant-current source is relatively unaffected by the magnetic circuit that it excites, except during transient conditions when changes are occurring in the magnetic circuit or in the external electric circuit. Under steady-state conditions with a constant-voltage source, the current in the coil is determined solely by the magnitude of the voltage source and the dc resistance of the coil.

In a circuit excited by a permanent magnet, the operating conditions of the permanent magnet are largely determined by the external magnetic circuit. Also, the operating point and subsequent performance of the permanent magnet are a function of how the magnet is physically installed in the circuit and whether it is magnetized before or after installation. In many applications, the magnet must go through a stabilizing routine before use. These considerations are, of course, meaningless for electrical excitation sources. For permanent magnet excitation, the object is to determine the size (length and cross section) of the permanent magnet. The first step in this process is to choose a specific type of permanent magnet, since each type of magnet has a unique characteristic that will partially determine the size of the magnet required. In a practical design this choice will be based on cost factors, availability, mechanical design (hardness and strength requirements), available space in the magnetic circuit, and the magnetic and electrical performance specifications of the circuit. Most permanent magnets are nonmachinable and usually must be used in the circuit as obtained from the manufacturer.

Permanent magnet excitation is chosen for a specified airgap flux density with the aid of the second-quadrant *B-H* curve, often called the *demagnetization curve*, for a specific type of magnet. This curve has been introduced in Fig. 1.4 (Section 1.2). The *B-H* characteristics of a number of Alnico permanent magnets are shown in Fig. 1.11. Also shown are curves of *energy product*, the product of *B* in gauss (G) and *H* in oersteds (Oe), and *permeance ratio*, the ratio of *B* to *H*. The energy product is a measure of the magnetic energy that the permanent magnet is capable of supplying to an external circuit as a function of its flux density and field intensity. In general, a permanent magnet is used most efficiently when operated at conditions of *B* and *H* that result in the maximum energy product. Permeance coefficients are useful in the design of the external magnetic circuit. This parameter is actually "relative permeability" as defined previously, since μ_0 is 1.0 in the CGS system of units. The symbols B_d for flux density and H_d for field intensity are used to designate the coordinates of the demagnetization curve.

Once the permanent magnet type has been chosen, the design of the magnet's size follows the general approach taken in Section 1.4. From Ampere's law,

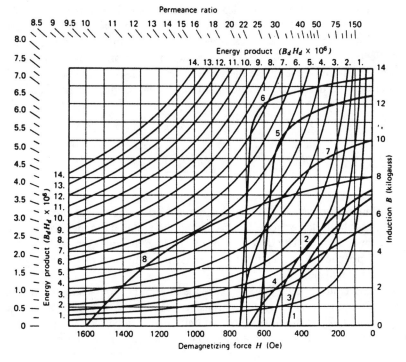

FIGURE 1.11
Demagnetization and energy product curves for Alnicos I to VIII. Key: 1, Alnico I; 2, Alnico II; 3, Alnico III; 4, Alnico IV; 5, Alnico V; 6, Alnico V-7; 7, Alnico VI; 8, Alnico VIII.

$$H_d l_m = H_g l_g + V_{mi} \tag{1.52}$$

where H_d = magnetic field intensity of the magnet, oe
$\quad l_m$ = length of magnet, cm
$\quad H_g$ = field intensity in the gap = flux density in gap, CGS units
$\quad l_g$ = length of gap, cm
$\quad V_{mi}$ = reluctance drop in other ferromagnetic portions of the circuit, G

The cross-sectional area of the magnet is calculated from the flux required in the airgap as follows:

$$B_d A_m = B_g A_g K_1 \tag{1.53}$$

where B_d = flux density in the magnet, G
$\quad A_m$ = cross-sectional area of magnet, cm^2
$\quad B_g$ = flux density in the gap, G
$\quad A_g$ = cross-sectional area of gap, cm^2
$\quad K_1$ = leakage factor

The leakage factor K_1 is the ratio of flux leaving the magnet to flux in the airgap. The difference between these two fluxes is the leakage flux in the regions of space between the magnet and the airgap. The leakage factor can be determined by the methods described in Sections 1.4 and 1.6 or by other standard and more accurate methods. Standard formulas are available in the literature to find the leakage factor.

It is interesting to determine the volume of permanent magnet material required to establish a given flux in an airgap. Solving for A_m in (1.53) and for l_m in (1.52) (neglecting V_{mi}) and noting that in the CGS system $H_g = B_g$, we obtain

$$\text{vol} = A_m l_m = \frac{B_g^2 A_g l_g K_1}{B_d H_d} \tag{1.54}$$

It is seen that magnet volume is a function of the square of the airgap flux density. The importance of the leakage factor in minimizing the required magnet size is also apparent from this equation. The denominator of (1.54) is the energy product that is a function of the permanent magnet material and the operating point on the demagnetization curve of the magnet.

The permeance ratio shown in Fig. 1.11 is the ratio of the equivalent permeance of the external circuit, $A_g K_1 / l_g$, to the permeance of the space occupied by the permanent magnet, A_m / l_m, in the CGS system of units. This can be seen by solving for B_d from (1.53) and for H_d from (1.52) (neglecting V_{mi}) and taking the ratio:

$$\frac{B_d}{H_d} = \frac{A_g l_m K_1}{A_m l_g} = \frac{P_{ge}}{P_M} = \tan\alpha \tag{1.55}$$

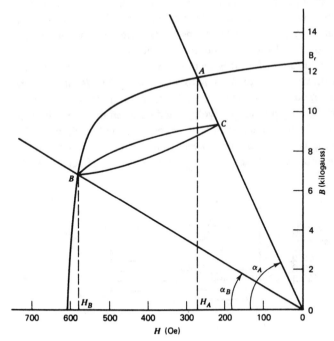

FIGURE 1.12
Second-quadrant B-H characteristic of a permanent magnet.

Equation (1.55) is deceptively simple in appearance because the task of obtaining analytical expressions for K_1 is very difficult. Also, the reluctance drop in the soft iron portions of the magnetic circuit, V_{mi}, must be included somehow in (1.55). This is an even more difficult task, since the reluctance drop is a function of both the permanent magnet's operating point, B_dH_d, and the effects of leakage flux in the iron. The reluctance drop is usually introduced by means of a factor similar to the leakage factor and is based on measurements in practical circuit configurations. The various expressions that make up (1.55) are of value in observing general relationships among the magnetic parameters as the permeance of the external circuit is varied.

Circuits with a varying airgap can be briefly described with the aid of Fig. 1.12. Keeping (1.55) in mind, let us observe the variation of B and H of a permanent magnet as the external circuit permeance is varied. Figure 1.12 shows a typical second-quadrant B-H characteristic for a permanent magnet. Theoretically, it is possible to have infinite permeance in the external magnetic circuit that would correspond to $\alpha = 90°$ in (1.55), and the magnet operating point would be at $B_d = B_r$ and $H_d = 0$ in Fig. 1.12. This situation is approximated by a permanent magnet having an external circuit consisting of no airgap and a high-permeability soft iron member, often called a "keeper." In practice, however, there is always a small equivalent airgap and a small reluctance drop in the keeper, and the operating point is to the left of B_r and α is less than $90°$.

For a finite airgap, the operating point will be at some point A on the B-H curve, and the permanent magnet will develop the magnetic field intensity H_A to overcome the reluctance drop of the airgap and other portions of the external magnetic circuit. If the airgap is increased, P_{ge} decreases, and the magnet must develop a larger magnetic field intensity H_d. From (1.55), it is seen that α decreases and the operating point on Fig. 1.12 will move farther to the left to some point, say B, at α_B. If the airgap is subsequently returned to its original value, the operating point will return not to A but, instead, to C. If the airgap is successively varied between the two values, the operating point will trace a "minor hysteresis loop" between B and C, as shown in Fig. 1.12. The slope of this loop is known as *recoil permeability*; since it is a slope on the B-H plane, it is also sometimes called incremental permeability, as defined in (1.26).

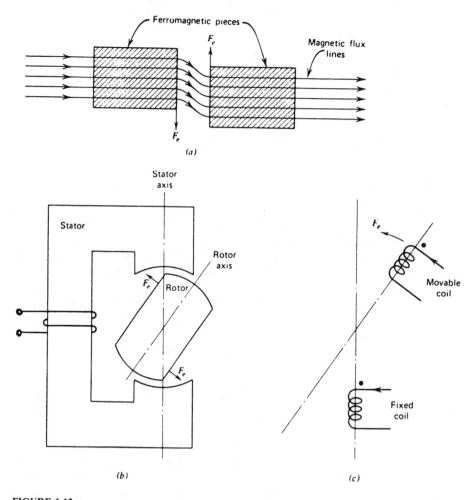

FIGURE 1.13

Magnetic field effects producing electrical force F_e. (a) Alignment of ferromagnetic pieces in a magnetic field; (b) a reluctance motor; (c) alignment of two current-carrying coils.

Recoil permeability is an important parameter of a permanent magnet for applications in magnetic circuits with varying airgaps.

1.10 FORCES OF ELECTROMAGNETIC ORIGIN

Up to this point we have considered only static magnetic circuits. However, for electromechanical energy conversion we must have mechanical motion and inherent forces to produce this motion (as in an actuator). The two basic magnetic field effects resulting in the production of mechanical forces are (1) alignment of flux lines and (2) interaction between magnetic fields and current-carrying conductors. Examples of "alignment" are shown in Fig. 1.13. In Fig. 1.13(a) the force on the ferromagnetic pieces causes them to align with the flux lines, thus shortening the magnetic flux path and reducing the reluctance. Figure 1.13(b) shows a simplified form of a reluctance motor in which electrical force tends to align the rotor axis with that of the stator. Figure 1.13(c) shows the alignment of two current-carrying coils. A few examples of "interaction" are shown in Fig. 1.14, in which current-carrying conductors experience mechanical forces when placed in magnetic fields. For instance, in Fig. 1.14(b) a force is produced by the interaction between the flux lines and coil current, resulting in a torque on the moving coil.

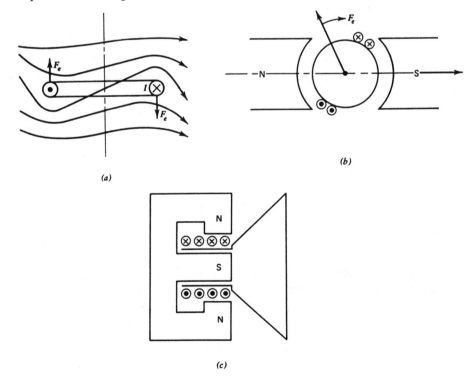

FIGURE 1.14
Electrical force produced by interaction of current-carrying conductors and magnetic fields. (a) A one-turn coil in a magnetic field; (b) a permanent magnet moving coil ammeter; (c) a moving coil loudspeaker.

1.11 THE FORCE EQUATION

We have given a few examples showing how mechanical forces are produced by magnetic fields. Clearly, for energy conversion (i.e., for doing work), mechanical motion is as important as mechanical force. Thus, during mechanical motion, the energy stored in the coupling magnetic field is disturbed. In Fig. 1.14(b), for instance, most of the magnetic field energy is stored in the airgap separating the rotor from the stator. The airgap field may be termed the *coupling field*. Electromechanical energy conversion occurs when coupling fields are disturbed in such a way that the energy stored in the fields changes with mechanical motion. A justification of this statement is possible from energy conservation principles, which will enable us to determine the magnitudes of mechanical forces arising from magnetic field effects.

Considering only the conservative portion of a system, such as a reluctance type actuator, we may write

$$
\begin{array}{ccccc}
\text{input} & & \text{mechanical} & & \text{increase} \\
\text{electrical} & = & \text{work} & + & \text{in stored} \\
\text{energy} & & \text{done} & & \text{energy}
\end{array}
\qquad (1.56)
$$

Or symbolically we have

$$
vi\,dt = F_e\,dx + dW_m
\qquad (1.57)
$$

where F_e is the force of electrical origin and dW_m is the increase in stored magnetic energy.

From Faraday's law, voltage v can be expressed in terms of the flux linkage λ as

$$
v = \frac{d\lambda}{dt}
\qquad (1.58)
$$

so that (1.57) becomes, after some rearrangement,

$$
F_e\,dx = -dW_m + i\,d\lambda
\qquad (1.59)
$$

In an electromechanical system either (i, x) or (λ, x) may be considered independent variables. If we consider (i, x) as independent, the flux linkage λ is given by $\lambda = \lambda(i, x)$, which can be expressed in terms of small changes as

$$
d\lambda = \frac{\partial\lambda}{\partial i}\,di + \frac{\partial\lambda}{\partial x}\,dx
\qquad (1.60)
$$

Also, we have $W_m = W_m(i, x)$, so that

$$dW_m = \frac{\partial W_m}{\partial i} \, di + \frac{\partial W_m}{\partial x} \, dx \tag{1.61}$$

Thus, (1.60) and (1.61), when substituted into (1.59), yield

$$F_e \, dx = \left(-\frac{\partial W_m}{\partial x} + i \, \frac{\partial \lambda}{\partial x} \right) dx + \left(-\frac{\partial W_m}{\partial i} + i \, \frac{\partial \lambda}{\partial i} \right) di \tag{1.62}$$

Because the incremental changes di and dx are arbitrary, F_e must be independent of these changes. Thus, for F_e to be independent of di, its coefficient in (1.62) must be zero. Consequently, (1.62) becomes

$$F_e = -\frac{\partial W_m}{\partial x}(i, x) + i \, \frac{\partial \lambda}{\partial x}(i, x) \tag{1.63}$$

which is the force equation and holds true if i is the independent variable.

If, on the other hand, λ is taken as the independent variable, that is, if $i = i(\lambda, x)$ and $W_m = W_m(\lambda, x)$, then

$$dW_m = \frac{\partial W_m}{\partial \lambda} \, d\lambda + \frac{\partial W_m}{\partial x} \, dx$$

which, when substituted into (1.59), gives

$$F_e \, dx = -\frac{\partial W_m}{\partial x} \, dx - \frac{\partial W_m}{\partial \lambda} \, d\lambda + i \, d\lambda \tag{1.64}$$

Because $\partial W_m / \partial \lambda = i$, (1.64) finally becomes

$$F_e = -\frac{\partial W_m}{\partial x}(\lambda, x) \tag{1.65}$$

The Concept of Coenergy and the Force Equation

Again considering the conservative portion of a system, let electrical energy be supplied to the input terminals and let there be no mechanical motion. In this case, the entire energy supplied is stored in the magnetic field, and this energy is

$$W_m = \int_0^\lambda i \, d\lambda'$$

This, when integrated by parts, yields

$$W_m = i\lambda - \int_0^i \lambda' \, di \qquad (1.66)$$

The quantity $\int_0^i \lambda' \, di$ in (1.66) is called *magnetic coenergy* W_m'. Thus, (1.66) can be written as

$$i\lambda = W_m + W_m' = \int_0^\lambda i \, d\lambda' + \int_0^i \lambda' \, di \qquad (1.67)$$

For a nonlinear magnetic circuit, W_m and W_m' are graphically depicted in Fig. 1.15. Clearly, for a linear magnetic circuit, magnetic energy and coenergy are equal. Following the procedure in deriving the force equations (1.63) and (1.65), it can be shown that, in terms of coenergy, the force equations become

$$F_e = \frac{\partial W_m'}{\partial x}(i, x) \qquad (1.68)$$

$$F_e = \frac{\partial W_m'}{\partial x}(\lambda, x) - \lambda \frac{\partial i}{\partial x}(\lambda, x) \qquad (1.69)$$

We now consider an example illustrating the application of the theory developed in this chapter.

Example 1.2 A solenoid of cylindrical geometry is shown in Fig. 1.16. (a) If the exciting coil carries a dc steady current I, derive an expression for the force on the plunger. (b) For the numerical values $I = 10$ A, $N = 500$ turns, $g = 5$ mm, $a = 20$ mm, $b = 2$ mm, and $l = 40$ mm, what is the magnitude of the force? Assume $\mu_{core} = \infty$ and neglect leakage.

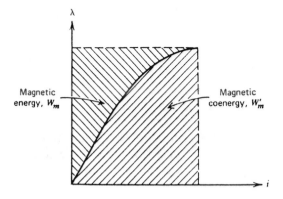

FIGURE 1.15
Magnetic energy and coenergy.

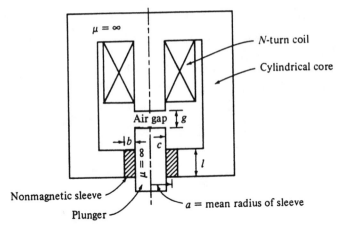

FIGURE 1.16
A solenoid of cylindrical geometry.

Solution

For the magnetic circuit, the reluctance is

$$\Re = \frac{g}{\mu_0 \pi c^2} + \frac{b}{\mu_0 2\pi a l} \quad \text{where} \quad c = a - \frac{b}{2}$$

The inductance L is then given by

$$L = \frac{N^2}{\Re} = \frac{2\pi\mu_0 a l c^2 N^2}{2alg + bc^2} \equiv \frac{k_1}{k_2 g + k_3}$$

where $k_1 \equiv 2\pi\mu_0 a l c^2 N^2$, $k_2 \equiv 2al$, and $k_3 \equiv bc^2$.

(a) Substituting $W_m = Li^2/2$ and $\lambda = Li$ in (1.62) yields

$$F_e = I^2/2 \frac{\partial L}{\partial g} = -\frac{I^2 k_1 k_2}{2(k_2 g + k_3)^2}$$

where the minus sign indicates that the force tends to decrease the airgap.

(b) Substituting the numerical values in the force expression of (a) yields 600 N as the magnitude of F_e, the force of electrical origin.

The background material developed in this chapter will aid our study of various linear actuator configurations in the following chapters.

REFERENCES

1.   S. A. Nasar, *Electric machines and power systems, Vol. I: Electric machines* (McGraw-Hill, New York, 1995).
2.   H. C. Roters, *Electromagnetic devices*, (Wiley, New York, 1941).

CHAPTER 2

INTRODUCTION TO LINEAR ELECTRIC ACTUATORS AND GENERATORS

In this chapter we present a general introduction to linear electric actuators (LEAs) and linear electric generators (LEGs). We will use the acronyms LEAs and LEGs consistently throughout the text. We begin with the definitions of LEAs and LEGs, although these definitions are not standardized in the profession. Subsequently, we will present a general discussion pertaining to the principles of operation and certain applications (existing and potential) of LEAs and LEGs.

2.1 TERMINOLOGY

Linear electric actuators (LEAs) are electromechanical devices which produce unidirectional or bidirectional short-stroke (less than a few meters) motion. In this regard, an LEA may be considered an electric motor in that both LEAs and electric motors internally develop electrical forces resulting in mechanical motion. However, the LEAs considered in this book have linear motion (or motion in a straight line), whereas electric motors have rotary motion. Moreover, unlike most electric motors, LEAs are commonly used for bidirectional, or reciprocating, motion.

Linear electric generators are electromechanical energy converters driven by prime movers undergoing reciprocating motion and converting mechanical power into single-phase ac power. It is to be noted that LEGs are not used for three-phase power

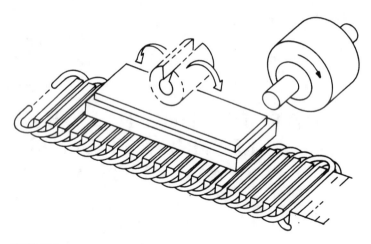

FIGURE 2.1
Unrolling a rotary electric machine.

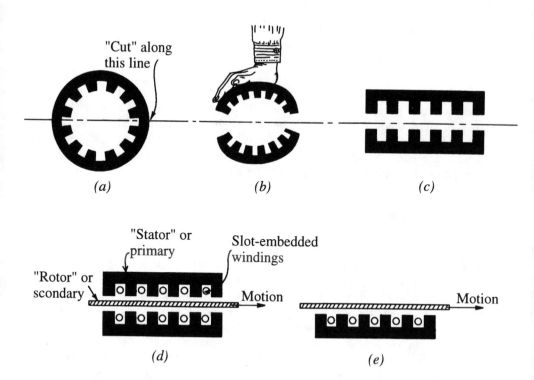

FIGURE 2.2
Halving and squashing a rotary electric machine.

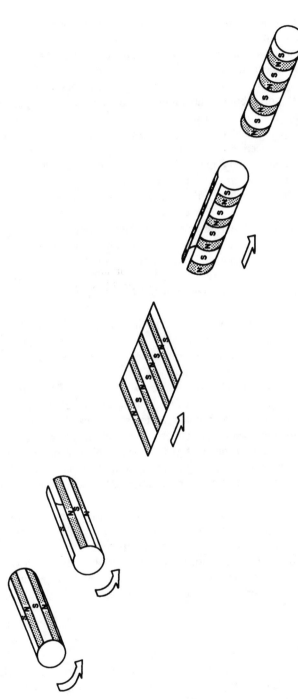

FIGURE 2.3
Rerolling to obtain tubular linear configurations.

generation, since the phase sequence of the generated voltage will reverse as the prime mover reverses its direction of motion.

Just as rotary electromagnetic machines do, LEAs and LEGs share the blessing of reversibility; that is, the same machine can act as an actuator or as a generator. However, in view of the single-phase output of LEGs, only single-phase LEAs used as short-stroke (up to 20-30 mm) vibrators have identical topologies. For longer (up to a few meters) travel, LEAs are in general three-phase machines [1-7] fed through power electronic controllers and taking advantage of regenerative electric braking for fast, robust, and precise thrust, speed, or position control.

LEAs and LEGs develop directly magnetic forces between a stator and a translator, or mover, acting upon a linear prime mover (for generating) or on a mover machine (for actuation). No mechanical transmission is required between the translator and the prime mover. Since there is a linear counterpart for every rotary electro-magnetic machine, we expect to encounter numerous conventional or easy to visualize LEAs and LEGs configurations. Figure 2.1 illustrates the now classic birth of the linear counterpart of a rotary induction machine by simply unrolling it. A *flat single-sided* topology is thus obtained. Alternatively, if the stator and rotor of a rotary machine are halved and "squashed" subsequently, a flat double-sided configuration results, as shown in Fig. 2.2.

The linear flat configuration may be used as is, or it may be rerolled along the direction of motion to form a tubular linear topology (Fig. 2.3). As the magnetic circuits in Figs. 2.2 and 2.3 become an open structure for the magnetic flux lines located in a plane containing the direction of motion, windings, field distributions, and performances are expected to show notable differences with respect to their rotary machine counterparts. However, in LEAs and LEGs, owing to the relatively short travel length, the mechanical airgap (between stator and translator) may be kept below 1 mm for most cases. Consequently the energy conversion performance is rather high, though we deal with low average speed (below 3 m/s) devices. This is especially true for devices utilizing high-energy permanent magnets to produce the magnetic fields. This is in contrast to linear electric motors for transportation [2], which are burdened by rather high, mechanically imposed airgaps (higher than 10 mm, in general). Single-sided flat LEAs and LEGs experience, besides the thrust force (along the direction of motion), a nonzero normal force (attraction or repulsion type) worthy of consideration for design purposes since the normal force may be greater than the thrust.

The above analogy points out some major similarities and differences between rotary electric machines and LEAs and LEGs. Let us now discuss the nature of developed force and the principles of operation of these devices.

2.2 OPERATION OF LEAs

Forces in magnetic fields may be either *electromagnetic* (by virtue of attraction, field alignment) or *electrodynamic* (Lorentz type). Electromagnetic forces occur by virtue of the principle of field alignment so as to reduce the magnetic stored energy. An elementary flat double-sided configuration with a ferromagnetic translator, as shown in Fig 2.4, illustrates the field alignment principle for a linear actuator operating as a

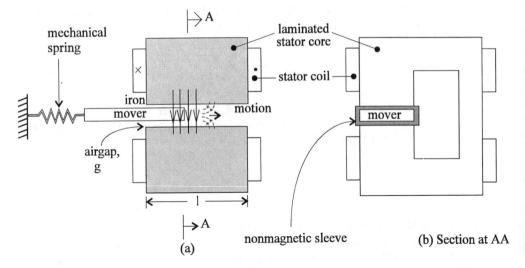

FIGURE 2.4
Field-alignment electromagnetic force in LEAs.

vibrator. When the stator coil is fed with an electric current, the ferromagnetic translator is pulled into the airgap (between the stator poles) until, for $x = 0$, the force becomes zero. At $x = 0$ all the flux lines ideally become vertical (aligned flux position) and the stored energy in the magnetic field is minimal. Now, if the current is turned off, the mechanical spring pulls the translator out from the airgap to $x = -l/2$ position. At this position the current is turned on again. Sustained oscillations may be obtained for resonance conditions. As presented in Chapter 1 and repeated here, the electromagnetic force, or the force of electrical origin F_e, is given by

$$F_e = \left(\frac{\partial W_m}{\partial x} \right)_{i=const} = \frac{1}{2} i^2 \frac{\partial L}{\partial x} \quad \text{N} \qquad (2.1)$$

where L is the inductance of the coil and i is the coil current.

As can be seen from Fig. 2.4, L increases with x. Therefore, when the translator moves from left to right, the force acts along the direction of motion. It is a thrust force irrespective of the current polarity. Similarly, if the current is nonzero when the translator moves from right to left, the electromagnetic force F_e maintains its direction but now acts against the direction of motion. A braking force is developed. To obtain bidirectional electromagnetic forces, more involved configurations have to be used but the principle still holds. Solenoids are a classic example of electromagnetic (attraction) force linear actuators.

Next, considering the Lorentz force as governed by the principle of electro-dynamic interaction, we observe that such a force can manifest in a number of ways. This force is the same as that often given by Ampere's force law, according to which a current-carrying conductor located in a magnetic field experiences a force. As discussed in Chapter 1, Ampere's force law is expressed as

$$\mathbf{F}_e = \mathbf{l}_a \, I \times \mathbf{B} \quad N \qquad (2.2)$$

where $\mathbf{F}_e$ is developed force, N; I is current, A; $\mathbf{l}_a$ is active length of the conductor, m; and $\mathbf{B}$ is magnetic flux density, T. On the other hand, in terms of force density f_e (N/m³), we have the Lorentz force equation:

$$\mathbf{f}_e = \mathbf{J} \times \mathbf{B} \quad N/m^3 \qquad (2.3)$$

where $\mathbf{J}$ is current density, A/m². To demonstrate the existence of electrodynamic forces, we refer to Fig 2.5, where the stator structure of Fig 2.4 is retained. The mover may be in the form of a coil or a conducting plate. The stator coil produces a magnetic field [$\mathbf{B}$ in (2.2) or (2.3)]. If this field is dc, the coil must be fed with direct current i from an external source. On the other hand, i may be induced (in the plate), in which case the stator coil is ac-fed. According to Lenz's law the flux of the induced current will oppose stator flux variation, and thus we have the polarity shown in Fig. 2.5(a). In an induction device the electrodynamic force will always tend to expel the coil (or the conducting plate) from the airgap. It is a force of repulsion. Again a spring is required to bring the mover back into the airgap at the $x = 0$ position with the stator coil current turned off during this stage. Sustained oscillations may for resonance

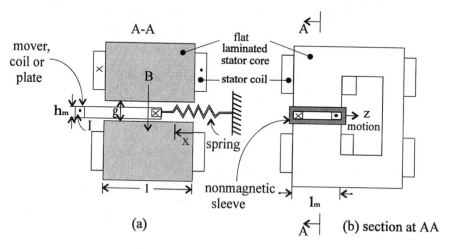

FIGURE 2.5
Lorenz (electrodynamic) force in LEAs.

conditions, resulting in an induction-type linear actuator operating as a vibrator. On the other hand, it is possible to have permanent magnets on the stator, instead of the stator coil, to produce a homopolar constant field in the airgap, with the mover coil fed from an ac source through a flexible electrical cable. In this case, the force changes direction with the mover current polarity and thus a push-pull type of force is produced. This is in fact the classic loudspeaker "voice coil" principle. A spring may be added at both stroke ends to improve the energy conversion efficiency, though the device itself is now, in principle, capable of sustained oscillations.

2.3 OPERATION OF LEGS

Like that of its rotary counterpart, the operation of an LEA is reversible; that is, an LEA could be made to operate as an LEG. Thus, the mover of an LEA may be actuated back and forth by a prime mover. (*Note*: Stirling engines and pneumatic, or air, motors have been used as prime movers to drive LEGs.) An ac voltage is motion-induced in the mover coil by the permanent magnet (PM) stator field, and an LEG emerges. An LEG with a PM mover can also be realized. A high-energy PM may be replaced by a fictitious air coil with a constant dc mmf I_{PM} such that

$$I_{PM} = H_c h_m \qquad (2.4)$$

where H_c is the PM coercive force and h_m is the PM height along the airgap g (Fig. 2.5(a)).

This reasoning also applies to the PM mover. Push-pull (bidirectional) force is obtained if the PM mover is actuated along the x-direction by a prime mover. The flux in the stator coil varies with the mover position from maximum to minimum. An ac motion-induced voltage is produced in the stator coil as a result of back-and-forth oscillatory motion of the PM translator. For flux reversal in the stator coil more involved configurations are required but the principle is basically the same.

In summary, the magnetic force classification leads to only two basic types of LEAs and LEGs: (a) those with passive ferromagnetic mover and stator coils and, eventually, PMs on the stator and (b) those with active diamagnetic mover (with air coils or PMs) and stator coils as well. This classification works for LEAs that work as vibrators and for LEAs and LEGs characterized by short stroke lengths (less then 20-30 mm, in general). For travels greater than 20-30 mm and up to a few meters, multiphase configurations as counterparts of rotary machines are used to provide low-ripple high-thrust levels required for the purpose. Such LEAs are hardly used as generators but regenerative braking is used for speed, position, or thrust control through power electronics. We may thus conceive linear induction, PM synchronous, reluctance synchronous, switched reluctance, and stepper actuators as linear counterparts of rotary machines.

2.4   EXISTING AND POTENTIAL APPLICATIONS

Linear induction actuators have been applied to moving machine-tool tables over a limited travel of 2 m or so. Thrust of about 1 kN for maximum speed of 2 m/s and copper losses of 0.8 W/N of thrust have been claimed with a cage stator and a 1-mm airgap. For the same purpose, linear PM synchronous actuators have also been designed [5]. This is a double-sided flat configuration with PMs placed along the entire travel length. This actuator has a mover similar to that of the linear induction actuator but is double-sided in the present case. Such linear PM synchronous actuators have been built for thrusts up to 20 kN with less than 0.8 W/N power/thrust ratio. The mover is directly attached to the work machine and thus no linear bearing is required. The airgap is kept around 1 mm to limit the PM weight. Power electronics, rectangular or sinusoidal, (vector) current control is used for thrust, speed, or position control. Only a small fraction of the PMs is active at any given time.

A two-phase hybrid PM reluctance actuator, commercially known as the Sawyer motor, is shown in Fig. 2.6 [6]. It has been used for linear positioning in computer peripherals and in sewing machines. Switched reluctance three-phase linear actuators [7,8] have been proposed for high-thrust-density applications in hostile environments such as in aircraft engine throttle positioning (Fig. 2.7). Complete decoupling of the three phases leads to a high reliability also. The three phases are turned on successively according to the ferromagnetic mover position through a unipolar current three-phase dc-dc power converter (or chopper).

Solenoid linear actuators [9] working on the electromagnetic force principle and voice coil actuators or loudspeakers [10] acting on the electrodynamic (Lorentz) force principle are designed on the basis of mature technologies with worldwide markets. Linear PM vibrators with tubular configurations and ac single-coil stator have been proposed for small linear pumps or air compressors [11]. The device uses mechanical springs at both ends of the stroke. Owing to the large weight of the mover the frequency of the sustained oscillations is limited to 5 Hz [11]. A 35-W rather high-frequency (120-Hz) vibrator with conduction cylinder light weight mover and unilat-

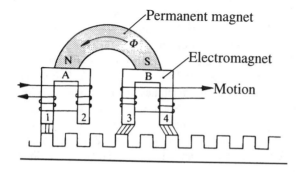

FIGURE 2.6
Linear two-phase hybrid PM reluctance stepper (Sawyer motor).

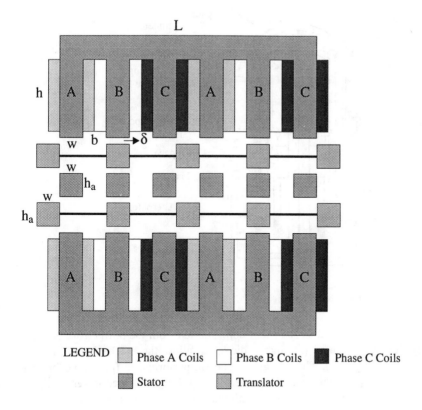

FIGURE 2.7
Linear three-phase switched reluctance actuator.

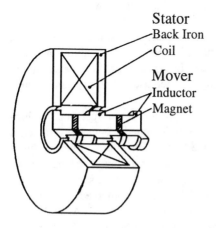

FIGURE 2.8
Linear conducting cylinder translator vibrator for pumping applications.

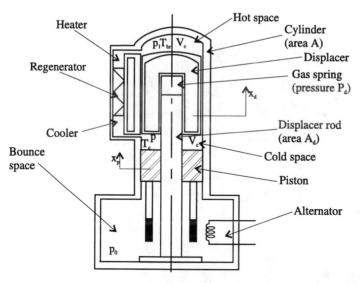

FIGURE 2.9
Linear PM translator alternator for Stirling engines.

eral-spring is shown in Fig. 2.8. The stator coil is ac-fed from the power grid, but one current polarity is eliminated (through a diode) to allow the mechanical spring to return the mover (Fig. 2.5). The energy conversion performance (power factor and efficiency) is rather poor and so is the power density, but the device is very simple, rugged, reliable, and thus suitable for actuating small-power linear pumps.

Free-piston Stirling engines have been proposed for space power applications or remote terrestrial areas for the conversion of low-temperature heat to mechanical energy. Linear single-phase tubular alternators have been matched with Stirling engines to produce electrical energy (Fig. 2.9) [12]. Linear alternators for powers up to 25 kW at industrial frequency and voltage (single-phase) at stroke lengths of up to 20-30 mm have been produced for good energy conversion performance by using PM movers and ring-shaped coils in the stator with series-connected capacitors for stability and PM weight reduction. Other configurations of LEAs and LEGs, with PMs on the stator and ferromagnetic movers, are under intense study to increase reliability and reduce fabrication and maintenance costs.

Linear multiphase actuators—induction, PM synchronous, or reluctance synchronous types [13]—have been proposed for magnetically levitated (contactless) material transfer in ultraclean (SCR wafer or optical device fabrication) rooms. Linear tubular PM synchronous actuators can be used for oil pumping or as low-frequency active shock absorbers in high-performance autos. As many linear position servos (used for locking car doors, tilting aircraft wings, automatic machine tool feeding, etc.) are becoming electric [14,15], we feel that the potential applications for LEAs and LEGs are enormous. The development of reasonably priced high-energy PMs, power electronics, sensors, and DSP controllers, along with good analysis and design tools should render the manufacture of LEAs and LEGs a rather independent new industry with worldwide markets.

REFERENCES

1. S. A. Nasar and I. Boldea, *Linear electric motors* (Prentice Hall, Englewood Cliffs, NJ, 1987).
2. I. Boldea and S. A. Nasar, *Linear motion electromagnetic systems* (Wiley-Interscience, New York, 1985).
3. G. W. Lean, "Review of recent progress in linear motors," *Proc. IEE*, vol. 135, pt. B, no. 6, 1988, pp. 380-416.
4. J. F. Eastham, "Novel synchronous machines: Linear and disc," *Proc. IEE*, vol. 137, pt. B, no. 1, 1990, pp. 49-58.
5. Anorad Corporation, *Anoline linear dc brushless servomotor* (Hauppage, NY).
6. T. Yokozuka and E. Baba. "Force displacement characteristics of linear stepping motors," *Proc. IEE*, vol. 139, pt. B, no. 1, 1992, pp. 37-43.
7. U. S. Deshpanday, J. J. Cathey, and E. Richter, "A high force density linear switched reluctance machine," *Rec. IEEE-IAS*, 1993 Annual Meeting.
8. J. Lucidarme, A. Amouri, and M. Poloujadoff, "Optimum design of longitudinal field variable reluctance motors: Application to high performance actuator," *IEEE Trans.*, vol. EC-8, no. 5, 1993.
9. Lucas Ledex, *Solenoid design guide* (1991).
10. B. Black, M. Lopez, and A. Marcos, "Basics of voice coil actuator," *PCIM Journal*, no. 7, July 1993, pp. 44-46.
11. J. Ebihara, H. Higashino, and T. Kasugai, "The optimum design of the cylindrical linear oscilatory actuator by analysis," *Rec. of ICEM*, 1990, pt. 2, pp. 527-531.
12. R. W. Redlich and D. W. Berchowitz, "Linear dynamics of free-piston Stirling engine," *Proc. I. Mech. Eng.*, March 1985, pp. 203-213.
13. M. Locci and I. Marongiu, "Modeling and testing of new linear reluctance motor," *Rec. ICEM*, 1992, vol. 2, pp. 706-710.
14. A. K. Budig, "Precision engineering demands for near linear drives in the micrometer and submicrometer range," *Rec. ICEM*, 1992, vol. 2, pp. 741-745.
15. H. Yamada, ed., *Handbook of linear motor applications*, 1986 (in Japanese).

LINEAR INDUCTION
ACTUATORS

Having given a general introduction to linear electric actuators in the preceding chapter, we now consider various types of LEAs in some detail. In this chapter, we begin with linear induction actuators (LIAs). These are essentially derived from rotary induction motors and resemble linear induction motors with restricted and controlled linear motion. Before we present an analysis of LIAs, it is worthwhile discussing some of the constraints under which these devices must operate.

3.1 PERFORMANCE SPECIFICATIONS

Linear induction actuators are used for short travel (between 2 and 3 m) and thus are characterized by rather small mechanical airgaps of about 1 mm. Also, the operating speed is generally less than 2 m/s. LIAs, as low-speed direct-drive systems, are characterized by the rated thrust F_{xm} for a given duty cycle and a peak (short-duration) thrust F_{xp} developed at standstill. LIAs are used only with variable-voltage variable-frequency static power converters or voltage-source inverters to obtain good energy conversion efficiencies and controlled thrust or positioning. The ac source—three-phase, in general—which supplies the static power converter is specified in terms of voltage and frequency: rated values and usual deviations. Few LIAs work continuously; short duty cycles are typical. In either case, the cooling system of the primary should also be specified as it determines the rated (design) current density in the LIA windings.

The performance of the control system of the LIA must also be specified. Finally, the LIA primary may use linear bearings, or may be attached directly to the work machine, or may be fully levitated to yield contactless mechanical linear motion. For travels up to 2 to 3 m the primary (with the three-phase ac windings) constitutes the mover. A flexible electric cable to supply the electric power to the primary is then required. Now, if the length of travel is considerly greater (up to 20 to 30 m), primary sections are placed along the track while the secondary (with a cage winding) acts as the mover. The primary sections are turned on only when the mover is in their proximity. On the other hand, tubular configurations are favored for travel lengths less than 0.5 m.

Because LIAs with long travel lengths are less commonly encountered, here we focus our attention on LIAs with moving primaries. As with any new power device, information on power factor and efficiency from the past is hardly relevant. However, thrust density ranges are known to be between 1 and 3 N/cm^2. In conjunction with allowable airgap flux densities and current density levels, they constitute an inadequate knowledge base for the design of LIAs. We now proceed with the construction details of LIAs.

3.2 CONSTRUCTION DETAILS FOR FLAT AND TUBULAR GEOMETRIES

An LIA may have a flat or a tubular configuration [1]. The flat configuration may be single-sided or double-sided. Conventional configurations have a conducting plate on a solid iron to form the secondary (Fig. 3.1) or may have only a conducting plate for the secondary (Fig. 3.2). However, for flat LIAs (with limited travel) secondaries with cage (or a short-circuited three-phase winding) placed in slots of a laminated core are used either in single-sided (Fig. 3.3) or in double-sided (Fig. 3.4) versions, to yield reasonably good energy efficiencies. Laminated cores on both sides, primary and secondary, result in low core losses and reasonably low magnetizing currents due to full use of back iron both in the primary and in the secondary. The slots in the primary and secondary may be open or semiclosed (Fig. 3.5). The open-slot geometry allows for preformed coils inserted in the slots. However, the airgap must be small (about 1 mm or less) to obtain a large airgap flux density. Because of open slots, the Carter coefficient K_c [1] is large. Carter's coefficient is given by

$$K_{c1,2} = \frac{1}{1 - \gamma_1 g/\tau_{s1,2}} \tag{3.1}$$

where

$$\gamma_{12} \approx \frac{(b_{so}/g)^2}{5 + b_{so}/g} \tag{3.2}$$

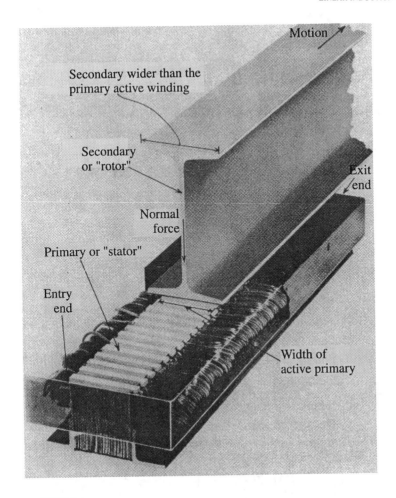

FIGURE 3.1
Flat LIA: single-sided, conventional.

$\tau_{s1,2}$ is the slot pitch of the primary or secondary, g is the airgap, and b_{so} is the slot opening. With both primary and secondary slotted, the airgap g will increase to an equivalent one g_e given by

$$g_e = K_{c1}K_{c2}g = K_c g \tag{3.3}$$

which depends essentially on the slot opening to airgap ratio and on the slot pitch.

As an example, for $\tau_s = 20$ mm, $g = 1$ mm, $b_{so} = 10$ mm for open slot and $b_{so} = 2.5$ mm for semiclosed slot, $K_c = 1.50$ and $K_c = 1.04342$, respectively. So open slots are used on one side only for high-thrust LIAs where coils are made of bars of transposed elementary rectangular cross-sectional conductors. For the tubular structure, open slots (Fig. 3.6) must be used for ease of manufacturing.

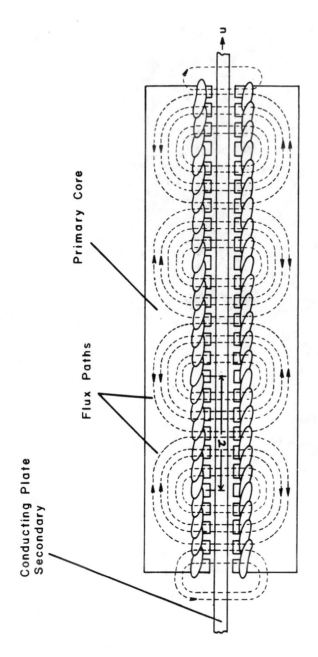

FIGURE 3.2
Flat LIA with a conducting plate secondary.

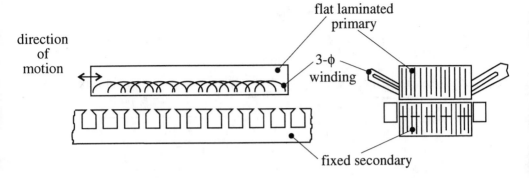

FIGURE 3.3
Flat LIA - single sided with cage secondary.

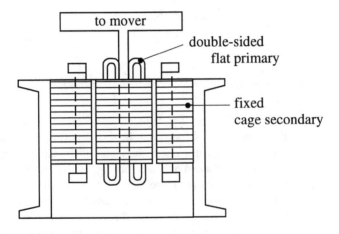

FIGURE 3.4
Flat LIA - double sided with cage secondary.

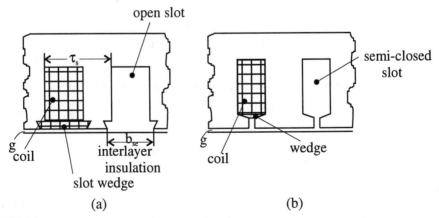

FIGURE 3.5
Typical practical slots (a) open and (b) semiclosed.

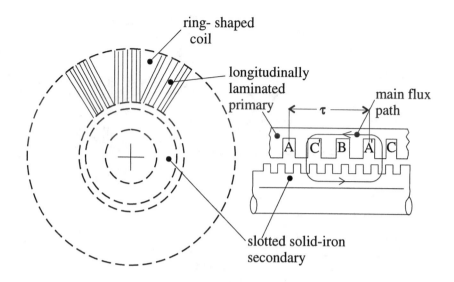

ring- shaped coil

longitudinally laminated primary

main flux path

slotted solid-iron secondary

FIGURE 3.6
Tubular LIA with longitudinal lamination primary.

The core of the primary of an LIA may be built either of longitudinal iron laminations (Fig. 3.6) or of disk (ring) shaped transverse laminations (Fig 3.7) with longitudinal laminations the magnetic flux paths are in the plane of the laminations. Thus, a good utilization of iron core is obtained and the core losses tend to be low. However, LIAs with longitudinal laminations are more difficult to manufacture. On the contrary, the transverse (disk) lamination core structures both for primary and secondary (Fig 3.7) are easy to fabricate. But in this case, the flux paths are through the laminations in the back iron core zone, resulting in an increase in the airgap per pole by the stacking factor $K_{iron} = 0.95$. Consequently,

$$g_{ad} \approx \frac{1}{3}\tau \left(1 - K_{iron}\right) \tag{3.4}$$

where τ = the pole pitch of the primary winding. The additional airgap occurs in both cores (primary and secondary) and the total airgap g_e' becomes:

$$g_e' = K_c g + 0.05 \left(\frac{2}{3}\tau\right)$$

For $g = 1$ mm, $K_c = 1.5$ and $\tau = 30$ mm, $g_e' = 1.5 + 1.0 = 2.5$ mm. A 1.0 mm increase in the airgap due to the stacking factor may be acceptable in view of savings in manufacturing costs. Also in the back cores of the primary and secondary the main

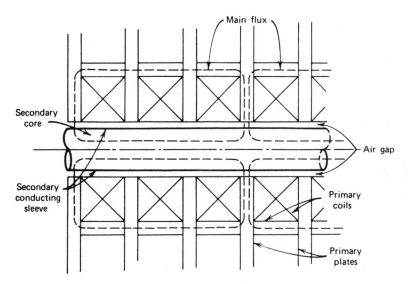

FIGURE 3.7
Tubular LIA with transverse disk laminations.

flux lines go through the plane of the laminations. Thus, large core losses due to the circular shape of eddy currents may result. To reduce these losses radial slits may be made in the back iron (Fig. 3.7). In general, six slits in the primary and four slits in the secondary are adequate to reduce the eddy current losses by a factor of about 6. Silicon steel laminations (with nonoriented grain) of 0.5 mm thickness are used for LIAs.

Turning now to the windings, generally three-phase windings are used, since three-phase inverters are commonly available. In principle, however, two-phase windings may also be used. LIA windings are very similar to those for rotary ac machines and ideally produce a traveling mmf. The magnetic circuit is open along the direction of motion. Consequently, a number of winding configurations for LIAs are feasible. Here we consider only four, which are of notable practical interest [1]:

- single-layer full-pitch ($y = \tau$) windings with an even number of poles $2p$ (Fig. 3.8) and one slot/pole-per phase ($q = 1$)
- triple-layer winding with an even number of poles $2p$ and $q = 1$ (Fig. 3.9) and $Y/\tau = 2/3$

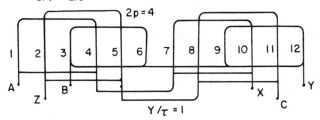

FIGURE 3.8
Single-layer full-pitch winding ($2p = 4$, $q = 1$).

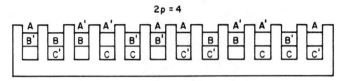

FIGURE 3.9
Triple-layer short-pitch ($y/\tau = 2/3$) winding ($2p = 4$, $q = 1$).

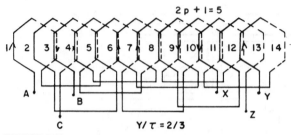

FIGURE 3.10
Double-layer short-pitch ($y/\tau = 2/3$) winding with $2p + 1$ poles and half-filled end slots.

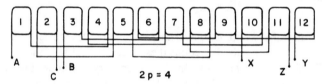

FIGURE 3.11
Economic single-layer winding ($2p = 4$).

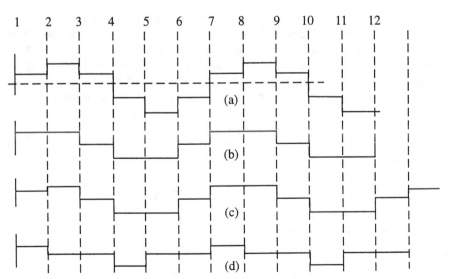

FIGURE 3.12
Magnetomotive force ideal distribution: $i_A = 1$; $i_B = i_C = -1/2$ for windings in Figs. 3.8(a), 3.9(b), 3.10(c), 3.11(d) for the same slot area.

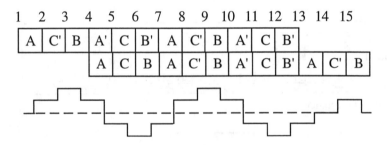

FIGURE 3.13
Double-layer full-pitch winding ($2p + 1 = 5$, $q = 1$): (a) layout; (b) mmf distribution when $I_A = 1$, $I_B = -1/2$, $I_C = -1/2$.

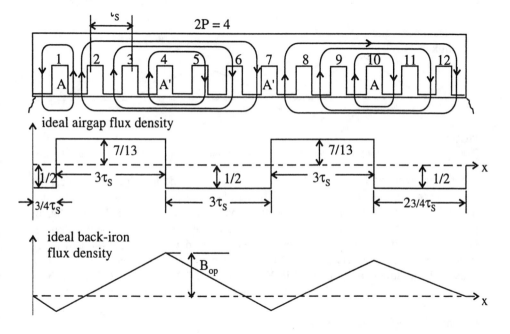

FIGURE 3.14
Phase A airgap and primary back-core flux densities for single-layer winding ($2p = 4$, $q = 1$).

- double-layer winding with odd number of poles $2p + 1$, $q = 1$ and half-filled end slots (Fig 3.10)
- "economic" windings for very low thrust LIAs (Fig. 3.11)

The mmfs of these windings, at the instant when $I_A = 1$, $I_B = I_C = -1/2$ are shown in Fig. 3.12. The mmf distribution shows that the maximum fundamental for a given slot (and current density) is obtained with a single-layer winding, the second in rank being the double-layer winding with an odd number of poles [Fig. 3.12(c)]. However, shortening the coils for $q = 1$ to $2/3$ produces a significant reduction in the fundamental. Thus, full-pitch coils should be used with one additional slot per primary. The

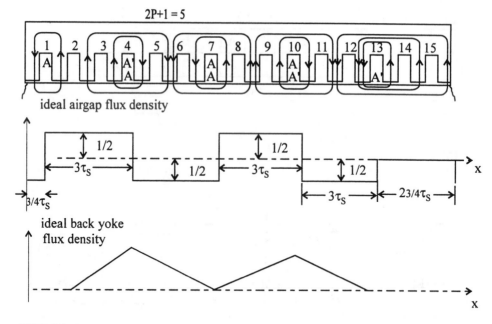

FIGURE 3.15
Phase ac airgap and primary back core flux densities for double-layer windings ($2p + 1 = 5$, $q = 1$).

"economic" winding has a deceptively low fundamental and is rich in harmonics, but it is easy to manufacture and, for miniature LIAs, such a winding is practical, since the end connections are short and contained within a small volume. For number of poles less than 5, the winding of Fig. 3.8 is preferred. For $2p > 5$ the full-pitch double-layer winding of Fig. 3.13 is applicable. Let us consider only one phase and apply Ampere's law and conservation of flux to determine the airgap and back iron flux density distributions, as illustrated in Figs. 3.14 and 3.15. It is evident from these figures that, based on magnetic permeance calculations, there is only a slight difference in the airgap and back-core flux density distribution for the two windings although in the end-pole region the difference is greater.

3.3 FIELD DISTRIBUTIONS

In the preceding section, we obtained the approximate flux density distribution by the magnetic permeance method with only one phase of the winding carrying current and with secondary current zero. Since q is 1, and at most could be 2, the airgap flux density distribution is far from the sinusoidal. Let us continue to explore this distribution by the magnetic permeance method for the two most practical windings, by adding the contributions of all three phases A, B, C with $I_A = +1$, $I_B = I_C = -1/2$ for zero secondary current. The results are shown in Figs. 3.16 and 3.17 for $2p = 4$ and $2p + 1 = 5$ and $q = 1$, respectively. The five-pole winding with half-filled end slots shows in fact four poles with one end pole having a lower flux density, whereas

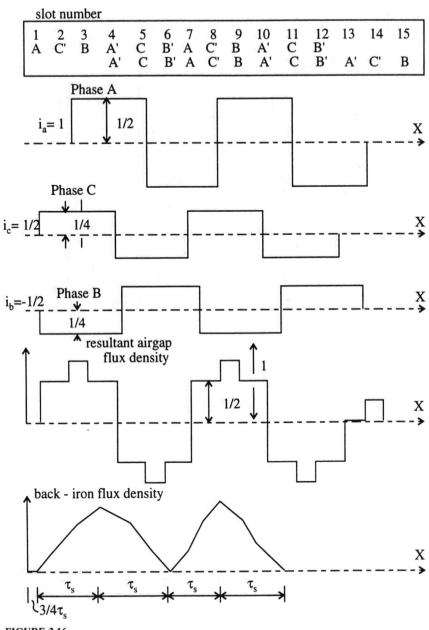

FIGURE 3.16
$2p = 4$ winding: resultant flux densities.

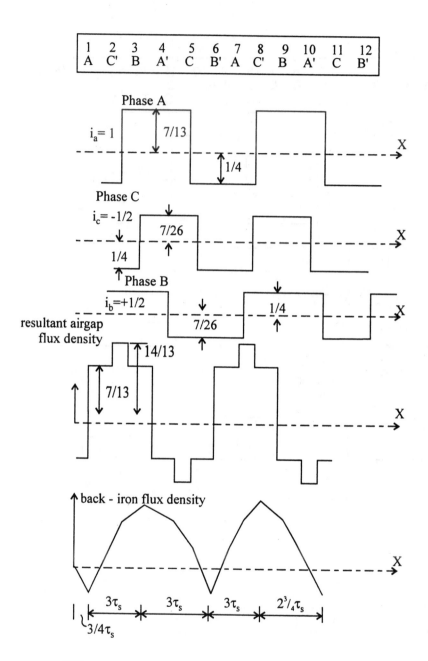

FIGURE 3.17
$2p + 1 = 5$ winding: resultant flux densities.

the four-pole winding uses the entire core to produce four almost symmetric poles. Notice that the total flux along the LIA length is zero. Finally, the flux density in the back iron is even smaller for the $2p$-pole winding, though only by a small amount. Thus, the difference between the two is only that one $(2p = 4)$ uses more copper for the end connections, whereas the other $(2p + 1 = 5)$ uses more iron. Also, the back-iron flux density adds up, in both cases, to almost the total pole flux and *not* half of it, as is the case in rotary electric machines. This is an important design aspect.

The presence of currents in the secondary for a nonzero relative speed between the two introduces new (zero current) zones of secondary under the primary and thus additional currents are produced in the secondary, in the regions of entry and exit ends of the primary [2-4]. Their influence on performance is called *dynamic longitudinal end effects*. It has been shown that the longitudinal end effects due to motion are small at the low speeds encountered in LIAs, despite the small number of poles [1]. Because the longitudinal end effect is small (the speed is only 2 m/s) it is apparent that the first pole is only slightly weaker than the rest of poles and that there are in all six strong poles, though the primary is seven poles long [8]. We may now draw the conclusion that for either of the two practical windings an equivalent single-layer $2p$-pole winding may be chosen when expressions for the flux density, thrust, and parameter expressions are investigated. In general, analytical solutions for the airgap flux density distributions are used for preliminary performance and design assessments with the finite-element method applied in the second stage for refinements.

The fundamental of the airgap flux density is of prime interest in the analysis of LIAs. For ideal no load (no current in the secondary), the fundamental of the airgap flux density B_{g10} is given by

$$B_{g10} = \frac{\mu_0 F_{10}}{g_e(1 + k_{so})}; \quad F_{10} = \frac{3\sqrt{2}\; N_1 K_{w1} I_{10}}{\pi p} \tag{3.5}$$

where F_{10} is the primary pole mmf fundamental, N_1 is turns per phase, K_{w1} is a winding factor (for $q = 1$ and $y = \tau$) p is pole pairs (either for $2p$ poles or for $(2p + 1)$ poles with half-filled end slots), $g_e = k_c g$ is the equivalent airgap, and k_{so} is a saturation factor which accounts for saturation in both the primary and secondary teeth and back-iron cores. The no-load characteristic can also be expressed as $B_{g10}(I_{10}/I_{1n})$; that is, the flux density as a function of the no-load current normalized by the primary rated current I_{1n}. For LIAs with an airgap of 1 mm (or less) the rated no-load airgap flux density is chosen just as in rotary induction motors; that is, $B_{g10} = 0.55 - 0.70$ T. Such values should also be the maximum encountered under peak short-duration thrust at standstill, to prevent high saturation levels.

In rotary induction machines there is an empirical relationship between primary diameter, airgap, and the number of poles. For LIAs the airgap g and the pole pitch τ should be closely related to yield good performance. The ratio τ/g should be as high as possible to reduce the magnetization current. But if the slots are open the Carter coefficient increases with the airgap decreasing and thus an adverse effect occurs. For open slots there is an optimum airgap, which, however, is not far from 1 mm. For semiclosed slots the airgap may be smaller and limited mechanically and corresponds to that of rotary machines for the same τ/g. For LIAs the pole pitch should not be too high to avoid too high slip frequencies for given thrust and to avoid thick back iron cores in the primary and secondary. It is felt that, in general, $\tau = (24$ to $60)$ mm for thrusts from a few newtons to a few kilonewtons.

The above discussion is intended to lead to a lumped parameter representation of LIAs similar to that applied to rotary induction machines, recognizing, however, the existence of net normal attraction forces in the single-sided LIAs. Such a normal attraction force exists also between the primary and each of the secondaries of double-sided LIAs. The normal force is an attraction force as the secondary has its cage (ladder for the flat configuration, ring-shaped conductors for the tubular configuration) placed in slots. Thus, the force of repulsion between primary and secondary mmfs, characteristic of linear induction motors with aluminum plate on iron secondary [1,4,7] is in fact negligible for the cage secondary.

3.4 LUMPED PARAMETER REPRESENTATION

Once the longitudinal end effect is neglected and the secondary is laminated and has a conducting cage in the slots, the LIA becomes similar to a rotary induction motor. Thus, the standard steady-state per-phase equations and equivalent circuits for steady-state typical of rotary induction motors may be applied. These equations are as follows:

$$I_1(R_1 + j\omega_1 L_{1\sigma}) - V_1 = E_1 \tag{3.6}$$

$$I_2'\left(\frac{R_2'}{s} + j\omega_1 L_{2\sigma}'\right) = E_1 \tag{3.7}$$

$$E_1 = -Z_{1m}I_{01} \; ; \qquad I_{01} = I_1 + I_2' \tag{3.8}$$

$$Z_{1on} = R_{1m} + j\omega_{1LIM} \tag{3.9}$$

where V_1 is the primary voltage per phase, E_1 is the main flux-induced voltage referred to the primary, R_2', $L_{2\sigma}'$ are secondary resistance and leakage inductance referred to the primary, I_1, I_2' are primary and secondary currents referred to the primary, R_{1m} is the core loss equivalent resistance per phase, L_{1m} is magnetizing inductance, ω_1 is primary frequency, and s is the slip, defined as

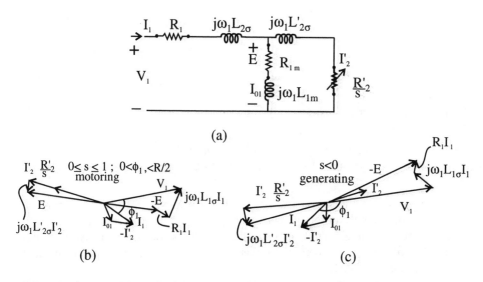

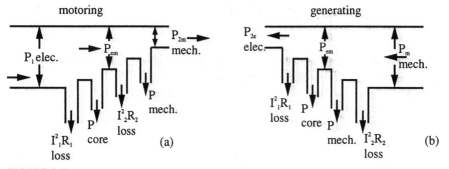

FIGURE 3.18
(a) Equivalent circuit and phasor diagram for (b) motoring and (c) generating.

FIGURE 3.19
Power flow for (a) motoring and (b) generating.

$$s = \frac{U_s - U}{U_s}; \qquad U_s = 2\tau f_1 \qquad (3.10)$$

In (3.10), τ is the pole pitch, f_1 the primary frequency, U_s the synchronous (ideal no-load) speed, and U the mover speed. The equivalent circuit and phasor diagrams (Fig. 3.18) are very similar to those of a rotary induction machine and are shown both for motoring and generating modes. The power flow and the losses are shown in Fig. 3.19 for both motoring and generating. Note that the electromagnetic power P_{em} is the product of electromagnetic thrust and linear synchronous speed U_s:

$$P_{em} = F_x U_s \qquad (3.11)$$

Furthermore,

$$P_{em} = 3 \frac{k_3}{s}(I_2')^2 \tag{3.12}$$

So the electromagnetic thrust F_x is given by

$$F_x = \frac{3R_2'(I_2')^2}{2\tau f_1 s} = \frac{3R_2'(I_2')^2}{2\tau f_2} \tag{3.13}$$

where f_2 is secondary frequency. The efficiency, power factor, and active and reactive powers are computed as for rotary induction machines. The normal attraction force F_{na} requires special treatment. First, as long as the airgap is constant no active power is associated with the normal force. So in this case it does not enter into the equivalent circuit or power flow diagram.

The attraction force between the primary and secondary iron cores is given by

$$F_{xa} = l \int_0^{2p\tau} \frac{B_{g1}^2(x,t)}{2\mu_0} \, dx = \frac{lB_{g1}^2}{4\mu_0} 2p\tau \tag{3.14}$$

where $2p\tau$ is the primary length along the direction of motion, l is the stack width, and B_{g1} is the amplitude of the resultant airgap flux density.

Using rotary machine theory it can be demonstrated that for an LIA the magnetizing inductance is

$$L_{1m} = \frac{6\mu_0(N_1 K_{w1})^2 \tau l}{\pi^2 p g_e(1 + k_s)} \tag{3.15}$$

The normal attraction force F_{na} is given by

$$F_{na} = \frac{3L_{1m}I_{01}^2}{g_e(1 + k_s)} \tag{3.16}$$

where I_{01} is the rms current. If there is linear motion in the direction of the airgap, at a speed $\dot{g}$ the corresponding active power P_{eg} becomes

$$P_{eg} = F_{na}\dot{g} \tag{3.17}$$

This effect can be introduced in the equivalent circuit by a resistor R_{eg} in series with R_{1m} and $j\omega_1 L_{1m}$, such that

$$R_{eg} = \frac{L_{1m}\dot{g}}{g_e(1 + k_s)} \tag{3.18}$$

Continuous vibrations along the airgap consume power as losses, and these losses are introduced in the equivalent circuit as R_{eg} in each phase.

3.5 STEADY-STATE CHARACTERISTICS

LIAs are likely to work only with static power converters. Thus, their steady-state characteristics are investigated for $V_1(f_1)$ relationships, typical for variable-frequency variable voltage control. The V_1/f_1 control is the most common in power electronics. To compensate for primary resistance voltage drop at low frequencies (or speeds) the $V_1(f_1)$ relationship is slightly modified to

$$V_1 = V_0 + k_f f_1 \tag{3.19}$$

where V_0 and k_f vary as desired, up to the maximum value of V_1/f_1, at which point the machine is fully saturated. The values of V_0 and k_f must be greater than a certain minimum for stability.

Referring to the equivalent circuit of Fig. 3.18, we have

$$I_2 \approx \frac{-V_1}{\sqrt{(R_1 + C_1 R_2'/s)^2 + \omega_1^2(L_{1\sigma} + C_1 L_{2\sigma}')^2}} \tag{3.20}$$

where $C_1 \approx 1 + L_{1\sigma}/L_{1m}$. Combining (3.13) and (3.20) yields

$$F_x = \frac{3V_1^2 R_2}{2\tau f_2[(R_1 + C_1 R_2'/s)^2 + \omega_1^2(L_{1\sigma} + C_1 L_{2\sigma}')^2]} \tag{3.21}$$

Similarly,

$$I_{01} = I_2 \sqrt{\frac{(R_2'/s)^2 + (\omega_1 L_{2\sigma}')^2}{R_{1m}^2 + (\omega_1 L_{1m})^2}} \tag{3.22}$$

Substituting (3.20) into (3.22) and the results into (3.16) yields the normal force (Fig. 3.16) as

$$F_{na} = \frac{3V_1^2 L_{1m}[(R_2'/s)^2 + (\omega_1 L_{2\sigma}')^2]}{g_e(1 + k_s)(R_{1m}^2 + \omega_1^2 L_{1m}^2)[(R_1 + C_1 R_2'/s)^2 + \omega_1^2(L_{1\sigma} + C_1 L_{2\sigma}')^2]} \tag{3.23}$$

The airgap flux density B_{g1} is given by

$$B_{g1} = \frac{3\sqrt{2}\,\mu_0 K_{w1} N_1 I_{01}}{\pi p g_e (1 + k_s)} \tag{3.24}$$

This value can be checked when the function $V_1(f_1)$ is modified. If severe saturation is avoided, by a proper choice of V_0 and k_f, even at low frequencies, the thrust/speed curves show a constant peak at different frequencies. This is accomplished by the control of the slip s_k as given by

$$s_k \simeq \frac{\pm C_1 R_2'}{\sqrt{R_1^2 + \omega_1^2 (L_{1\sigma} + L_{2\sigma}')^2}} \tag{3.25}$$

The value of I_{01} is expected to vary. Thus, the normal force varies considerably. Qualitatively, such characteristics are shown in Fig. 3.20. Notice that as U_s varies with frequency, the slip s is determined from

$$s = 1 - \frac{u}{2\tau f_1} \tag{3.26}$$

In Fig. 3.20, u_{sn} is the synchronous speed at the rated frequency. We observe that the thrust/speed curves exhibit one maximum for motoring and one for generating and thus have a limited static stability zone.

During transients especially at low speeds, instabilities may occur if the frequency f_1 is ramped faster than the actuator can follow it by speed response. To obtain linear thrust/speed curves and fast response, controlled flux is introduced as vector control. At this point we are interested only in the steady-state characteristics for controlled flux. The flux in a primary phase λ_1 is given by

$$\lambda_1 = L_{1\sigma} I_1 + L_{1m}(I_1 + I_2') \tag{3.27}$$

The magnetizing (airgap) flux λ_{1m} is expressed as

$$\lambda_{1m} = L_{1m}(I_1 + I_2') = L_{1m} I_{01m} \tag{3.28}$$

which is the main flux (also known as airgap flux). Finally, the secondary flux λ_2 referred to the primary is

$$\lambda_2' = L_{2\sigma}' I_2' + \lambda_{1m} \tag{3.29}$$

with

$$\lambda_1 = L_{1\sigma} I_1 + \lambda_{1m} \tag{3.30}$$

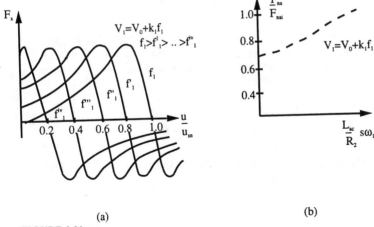

(a)　　　　　　　　　　　　(b)

FIGURE 3.20
Force/speed curves for $V_1 = K_0 + K_f f_1$: (a) thrust, (b) normal force.

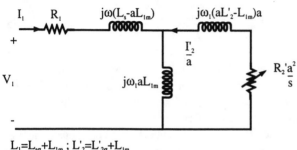

$$L_1 = L_{s\sigma} + L_{1m} \; ; \; L'_2 = L'_{2\sigma} + L_{1m}$$

FIGURE 3.21
General flux equivalent circuit (core loss neglected: $R_{1m} = 0$).

To control (or maintain a constant amplitude of) any of the quantities λ_1, λ_{1m}, λ_2, change in the variable I_s is introduced [9-10] such that

$$I'_{2a} = \frac{I'_2}{a} \tag{3.31}$$

When $a = 1$ the equivalent circuit of Fig. 3.18 is obtained; that is, the magnetizing (airgap) flux predominates. Keeping I_{01m} constant in amplitude implicitly means a constant airgap flux and flux density, and consequently a normal constant force. Using this change of variables in (3.6) through (3.9), the equivalent circuit of Fig. 3.21 [10] is obtained. This circuit is called the general flux equivalent circuit. For $a = 1$ we notice that λ_{0a} becomes

$$\lambda_{0a} = aL_{1m}I_{0a} = L_{1m}(I_1 + I'_2) = L_{1m}I_{0m} = \lambda_{1m} \tag{3.32}$$

Therefore, we now explore first the constant magnetizing flux λ_{1m} characteristics, with $|I_{01m}| = $ constant.

We observe that with a constant I_{01}, the normal attraction force F_{na}, as given by (3.16), is strictly constant for a constant airgap, as L_{1m} remains constant. However, the thrust as obtained from (3.21) and (3.23) with core losses neglected ($R_{1m} = 0$) becomes

$$F_x = \frac{3\pi R_2'}{\tau} \frac{\lambda_{1m}^2}{((R_2')^2/s\omega_1) + s\omega_1 L_{2\sigma}^2} \tag{3.33}$$

It follows from (3.33) that the thrust is proportional to the square of the airgap flux λ_{1m}^2, and is maximum for $s_{k0}\omega_1$ given by

$$s_{k0}\omega_1 = \pm\frac{R_a'}{L_{2\sigma}'} = \omega_{2k0} \tag{3.34}$$

The peak thrust F_{x0k} is:

$$F_{x0k} = \frac{3\pi}{2\tau} \frac{\lambda_{1m}^2}{L_{2\sigma}'} \tag{3.35}$$

The smaller the $L'_{2\sigma}$, the higher is the peak thrust, but the secondary frequency $\omega_{2k0} = s_{k0}\omega_1$ also becomes higher. For an aluminum sheet on iron secondary, $L'_{2\sigma}$ is really very small (ideally zero), and thus ω_{2k0} is large and so is F_{x0k}. This definite merit is offset, however, by a poor power factor and efficiency.

The mechanical characteristics for variable frequency and constant airgap flux are shown in Fig. 3.22. It should be noted that $s_{k0}\omega_1 > s_k\omega_1$. Therefore, the stability zone increases considerably. Also, the control of λ_{1m} provides direct control of normal (attraction) force for suspension control when so desired. Constant normal (attraction) force may also be desired to reduce vibration and noise even when only the thrust is controlled by frequency f_1 with λ_{1m} constant in amplitude.

Next we consider constant secondary flux characteristics. With $a = L_{1m}/L_2'$, Fig. 3.23 shows that the inductance in the secondary "disappears," only to be moved to the primary. However, in this case we have

$$\lambda_{0a} = aL_{1m}I_{0a} = \frac{L_{1m}}{L_2'} L_{1m} \left(I_1 + \frac{I_2'L_2'}{L_{1m}}\right) = \frac{L_{1m}}{L_2'} \lambda_2' \tag{3.36}$$

So, in fact, in this case $|\lambda_2'|$ is kept constant. From (3.27) $\underline{\lambda}_1$ becomes

$$\lambda_1 \approx L_{sc}i_1 + \frac{L_{1m}}{L_2'} \lambda_2' \tag{3.37}$$

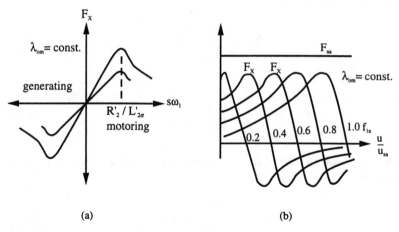

FIGURE 3.22
Constant airgap flux characteristics: (a) thrust versus secondary (slip) frequency; (b) F_x and normal attraction force F_{na} versus speed.

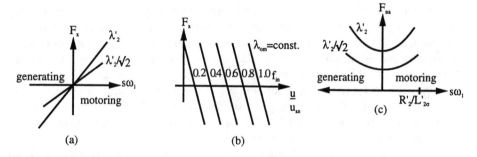

FIGURE 3.23
Constant secondary flux characteristics: (a) thrust versus secondary (slip) frequency; (b) thrust versus speed; (c) normal attraction force versus secondary frequency.

Therefore, the new leakage inductance in the primary circuit of Fig. 3.23 is the short-circuit inductance L_{sc} as given by

$$L_{SC} = L_1 - \frac{L_{1m}^2}{L_2'} \approx L_{1\sigma} + L_{2\sigma}' \tag{3.38}$$

The thrust can again be determined from (3.13), with

$$\lambda_2' \frac{L_{1m}}{L_2'} \omega_1 = \frac{I_2'}{L_{1m}/L_2'} \left(\frac{L_{1m}}{L_2'}\right) \frac{R_2'}{s} \tag{3.39}$$

or

$$I_2' = \frac{s\omega_1 \lambda_2'}{R_2'} \tag{3.40}$$

The thrust F_x is given by

$$F_x = \frac{3\pi}{\tau} \frac{(\lambda_2')^2}{R_2'} s\omega_1 \tag{3.41}$$

To obtain an expression for the normal force we use (3.23) with $R_{1m} = 0$. Consequently, we have

$$F_{na} = \frac{3(\lambda_2')^2}{L_{1m}g_e(1 + k_s)} \left[1 + \left(\frac{\omega_1 s L_{2\sigma}'}{R_2'}\right)^2\right] \tag{3.42}$$

For a constant level of saturation we have λ_2' = constant, L_{1m} = constant, and k_s = constant. Therefore, both the thrust and normal force depend on the secondary flux squared, $(\lambda_2')^2$, and do not show any extremes. The thrust varies linearly with secondary frequency, giving rise to linear mechanical characteristics such as those of separately excited dc machines (Fig 3.23). On the other hand, the normal force tends to increase with the secondary frequency. This aspect should be kept in mind when applying constant secondary flux control for variable frequency either for thrust control only or for normal attraction (suspension) force control. So in fact the airgap flux increases with $s\omega_1$ for constant λ_2'. To avoid saturation under heavy loads a lower value of airgap flux for zero thrust must be assigned.

"Moving" the secondary leakage inductance to the primary circuit produces fast thrust response with current control. The primary current I_1 has two components: one is the field current I_{0s} which produces the secondary flux λ_2', and the other is the thrust current i_T, as given by

$$I_{0s} = \frac{\lambda_2'}{L_{1m}}; \quad I_T = I_2' \frac{L_2'}{L_{1m}} = \frac{s\omega_1 \lambda_2'}{R_2'} \frac{L_2'}{L_{1m}} = \omega_1 s \frac{L_2'}{R_2'} I_{0s} \tag{3.43}$$

The thrust can then be expressed as

$$F_x = 3\frac{\pi}{\tau} \frac{L_{1m}^2}{L_2'} I_{0s} I_T \tag{3.44}$$

The primary current in terms of its components is

$$I_1 = I_T - jI_{0s} \tag{3.45}$$

At the primary frequency f_1, the currents $i^*_{a,b,c}(t)$ for the phases a, b, c, are

$$i^*_{abc}(t) = \sqrt{2}\,I_T \cos[\omega^*_1 t - \frac{2\pi}{3}(k-1)] + \sqrt{2}\,I_{0s} \sin[\omega^*_1 t - \frac{2\pi}{3}(k-1)]$$
$$k = 1,2,3 \tag{3.46}$$

As can be seen from the above, given the thrust F_x and secondary flux λ'_2 requirements (reference values), the value of $s\omega_1$ is first obtained from (3.43), with the speed U measured:

$$\omega_r = \frac{\pi}{\tau} U \tag{3.47}$$

The reference primary frequency ω_1^* becomes

$$\omega_1^* = \omega_r + (s\omega_1)^* \tag{3.48}$$

Then the reference phase currents $i^*_{abc}(t)$ can be calculated in a microcontroller and "reproduced" by using current sensors and current controllers through a PWM static power converter with high switching frequency. This is the counterpart of constant rotor flux (vector) control of rotary induction motors [10].

Finally, we consider constant primary flux characteristics. As implied by (3.40), any variation of secondary resistance with temperature would produce secondary frequencies $s\omega_1$, for given torque current I_T, different from those required. So, unless some R'_2 adaptation is performed, the good linear thrust speed characteristic is lost. On the other hand, a constant primary flux operation is also feasible. It is obtained for $a_p = (L_1/L_{1m})$ when the total leakage inductance in Fig. 3.21 is moved to the secondary and λ_{0a} becomes

$$\lambda_{0a} = a_p L_{1m} I_{0p} = \frac{L_1}{L_{1m}} L_{1m} I_{0p} = L_1 I_{0p} = \lambda_1 \tag{3.49}$$

and

$$I_{0p} = I_1 + I'_2 \frac{L_{1m}}{L_1} \tag{3.50}$$

From Fig. 3.20(a) we obtain

$$I'_2 = \frac{-j\omega_1 \lambda_1 (L_{1m}/L_1)}{j\omega_1 (L'_2 - (L^2_{1m}/L_1)) + R'_2/s} \tag{3.51}$$

Let us abbreviate

$$L'_2 - \frac{L^2_{1m}}{L_1} \approx \frac{L'_2}{L_1} L_{sc} = L'_{sc} \tag{3.52}$$

In general, $L_2' \approx L_1$ so $L_{sc}' \approx L_{sc}$. Therefore, the thrust from (3.13) becomes

$$F_x = \frac{3\pi}{\tau} R_2' \frac{\lambda_1^2(L_{1m}/L_1)^2}{s\omega_1(L_{sc}')^2 + (R_2')^2/s\omega_1} \tag{3.53}$$

The thrust has again a maximum (for positive and negative values) at

$$(s\omega_1)_{kp} = \pm \frac{R_2'}{L_{sc}'} \tag{3.54}$$

And the peak thrust F_{xp} assumes the form

$$F_{xp} = \frac{3\pi}{2\tau} \frac{\lambda_1^2(L_{1m}/L_1)^2}{L_{sc}'} \tag{3.55}$$

The normal force F_{na} is obtained by making use of (3.16), (3.23), and (3.52) to obtain

$$F_{na} = \frac{3}{g_e(1 + k_s)L_{1m}} \lambda_1^2 \frac{[1 + (s\omega_1 L_{2\sigma}'/R_2')^2]}{1 + (s\omega_1 L_{sc}'/R_2')^2} (L_{1m}/L_1)^2 \tag{3.56}$$

As $L_{2\sigma}' < L_{sc}'$ it is obvious that F_{na} decreases with $s\omega_1$. For example, for the peak thrust at the slip frequency given by (3.54) we have

$$F_{nap} = \frac{3}{L_{1m}g_e(1 + k_s)} \lambda_1^2 \frac{[1 + (L_{2\sigma}'/L_{sc}')^2]}{2} \left(\frac{L_m}{L_1}\right)^2 \tag{3.57}$$

Now as $L_{2\sigma}'/L_{se}' \leq 1/2$ the decrease in the normal force is significant with $s\omega_1$ (or load) increasing. If normal force control for suspension control is targeted, care must be exercised because of the rather involved load-dependent variation of normal force.

It should be noted that constant λ_1 (Fig. 3.24) and V_1/f_1 controls are similar under steady state. They both exhibit a rather narrow static stability region with notable variation in the normal force.

3.6 STATE-SPACE EQUATIONS

As we have already mentioned that LIAs behave like cage-rotor type of induction machines, the space-phasor equations [10] may be directly applied here, replacing the torque by thrust. Notice that the primary is considered the translator (mover). The translator (primary) moves in the opposite direction of the traveling field of the primary. The reference system is considered to move at an equivalent speed ω_b

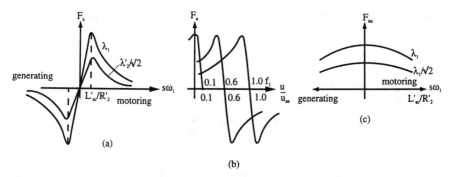

FIGURE 3.24
Constant primary flux characteristics: (a) thrust versus secondary frequency; (b) thrust versus speed; (c) normal forces versus secondary frequency.

considered positive along the direction of the traveling field (opposite to the mechanical notion of the primary). Thus,

$$\omega_b = \frac{\pi}{\tau} u_b \tag{3.58}$$

where u_b is the linear speed of the reference system. Noting this fact, the space-phasor equations of rotary induction machines [10] may be applied to obtain

$$v_1 = R_1 i_1 + \frac{d\lambda_1}{dt} + j\omega_b \lambda_1; \qquad v_1 = v_d + jv_q; \qquad i_1 = i_d + ji_q \tag{3.59}$$

$$0 = R_2' i_2' + \frac{d\lambda_2'}{dt} + j(\omega_b - \omega_r)\lambda_2'; \qquad i_2 = i_D + ji_Q; \qquad \lambda_i = \lambda_{di} + j\lambda_{qi} \tag{3.60}$$

$$F_x = \frac{-3\pi}{2\tau} \operatorname{Im}(\lambda_1 i_1^*); \qquad \lambda_1 = L_1 i_1 L_{1m} i_2'; \qquad \lambda_2' = L_2' i_2' + L_{1m} i_1 \tag{3.61}$$

In (3.61), F_x is the force exerted on the primary (translator). Now, Park transformation is used to relate the space-phasor quantities to phase quantities for the primary; that is,

$$v_1^b = \frac{2}{3}(v_a + v_b e^{j2\pi/3} + v_c e^{-j2\pi/3})e^{-j\theta_b}; \qquad \frac{d\theta_b}{dt} = \omega_b \tag{3.62}$$

The normal force is related directly to the amplitude of the airgap flux per phase, λ_{1m}. Thus,

$$F_{na} = \frac{3\lambda_{1m}^2}{2g_e(1 + k_s)L_{1m}} \tag{3.63}$$

The same amplitude of the flux holds for the space-phasor model; that is,

$$\lambda_{1m} = |\lambda_1 - L_{1\sigma}i_1| \tag{3.64}$$

Now there are four equations of motion:

$$M\frac{du}{dt} = F_x - F_{xload}; \qquad u = \frac{dx}{dt}; \qquad \frac{\pi}{\tau}u = \omega_r \tag{3.65}$$

$$M\frac{d\dot{g}}{dt} = F_{na} - Ma_g - F_{nload}; \qquad \dot{g} = \frac{dz}{dt} \tag{3.66}$$

where a_g is the acceleration due to gravity and M is the translator mass. Decomposing into d-q components, the space-phasor equations can be solved by numerical methods for transients as well as for steady state, when the sum of the primary currents is zero in a wye connection. The zero sequence component equations should otherwise be added [10]. In phase coordinates the state-space equations involve time-varying mutual inductances between primary and secondary. Consequently, their solutions require more computer time. However, if suspension control is applied, the airgap varies and so does L_{1m} and, through it, $L_1 = L_{1\sigma} + L_{1m}$ and $L_2' = L_{2\sigma}' + L_{1m}$. As a first approximation L_{1m} can be written as

$$L_{1m} = \frac{C_{1m}}{g} \tag{3.67}$$

As ω_b reverses, for $\omega_b = \omega_1$ (synchronous coordinates), we have

$$\theta_b^s = \int \omega_1 \, dt + \theta_{b0}^s = -\int (s\omega_1 + \omega_r) \, dt + \theta_{b0}^s \tag{3.68}$$

The variables can be the flux components λ_d, λ_q, λ_D, λ_Q, u, g, x, and z; θ_b^s, with the inputs v_d, v_q, F_{xload}, F_{nload}, ω_{10}. The reference speed or position along the direction of thrust motion, u^* or x^*, and the reference g^* (to be kept constant through normal force control) are produced by their respective controllers. For constant airgap (no suspension control), the state-space equations are greatly simplified.

3.7 VECTOR CONTROL ASPECTS

LIAs are used to produce linear motion with thrust or speed or position control at constant airgap. They are also used for combined propulsion and suspension systems. The simplest way to produce controlled propulsion positioning is to use V_1/f_1 control with a PWM voltage-source inverter, as is the case for most rotary induction motor variable speed drives. If proper limitations in $V_1(f_1)$ function are introduced a priori, the drive may be made stable and robust, and moderately quick in response to ensure stability. The controller operates on frequency and voltage amplitudes rather than on the phase shift of the primary voltage, resulting in a slower response. For quicker thrust and position control, vector control may be used. Secondary flux orientation vector control is adequate in terms of fast transients and good stability. Indirect vector current control [10] is widely used for rotary induction motors and is adaptable to LIAs.

As an application example, a direct primary flux λ_1 and thrust vector control system applied for both propulsion and suspension simultaneous control is presented in the following.

Direct Vector Control of LIA Propulsion and Suspension

Transfer machines are necessary in ultraclean environments. For such applications LIAs with simultaneous propulsion and suspension control are quite suitable. We recall that the space-phasor primary equation (3.59) in primary coordinates ($\omega_b = 0$) becomes

$$v_1 = R_1 i_1 + \frac{d\lambda_1}{dt} \tag{3.69}$$

Integrating with respect to time yields

$$\lambda_1 - \lambda_{10} = \int_0^t \left[v_1(t) - R_1 i_1 \right] dt \tag{3.70}$$

For control purposes we may neglect R_1 to obtain

$$\lambda_1(t) - \lambda_1(0) \approx \int_0^t v_1(t) \, dt \tag{3.71}$$

PWM voltage-source inverters (Fig. 3.25) provide for six nonzero voltage space phasors, depending on which three SCRs are conducting at a time, and two zero voltage vectors when all three upper- or lower-leg SCRs are conducting. The voltage vectors v_0 through v_6 and the phase voltages for V_0 dc link voltage are given in Table 3.1.

Table 3.1 is obtained by observing that at any time $V_a + V_c + V_c = 0$ (phases being wye connected) and that one phase is in series with the other two in parallel for each distinct voltage vector. Notice from (3.70) that the primary flux variation falls along the voltage vectors $v(i)$ currently produced by the inverter. For clockwise motion of the traveling field and the flux vector in Sector 4 [$\theta(1) = \pm 30°$ around the axis of phase a], we may increase the flux by turning on either V_4 or V_6, and for decreasing it, V_2 is to be used. Increasing the primary flux means in fact also increasing the airgap flux and thus normal attraction (suspension) force is also increased.

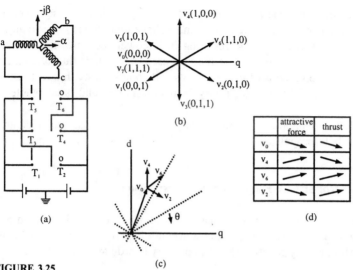

FIGURE 3.25
Direct primary flux control: (a) PWM voltage inverter with ideal switches; (b) available voltage vectors; (c) changing the thrust and normal force for the flux vector the first sector ($\theta(1)$) ±30° around phase a axis.

To produce propulsion we must accelerate the field along the direction of motion and thus use V_6 or V_2. To produce a braking force we decelerate the motion of the flux vector by applying V_4. If both thrust and normal forces are to be reduced V_0 is turned on [Fig. 3.25(c)] [11]. The motion may be forward or backward (f/b), and when the thrust error digital signal τ, the flux error signal λ_1, and the flux vector current position sector (θ_i) are added a table of adequate switchings in the inverter is obtained (Fig. 3.26) [11]. We need, however, the thrust and flux errors. The reference thrust F_x^* and reference flux λ_1^* should be outputed by the propulsion and suspension controller. The primary flux amplitude λ_1 is obtained from (3.70) or (3.71) after using Park transformation for $\theta_b = 0$ but for a clockwise motion:

$$V_1 = \frac{2}{3}\left(V_a + V_b e^{-j2\pi/3} + V_c e^{j2\pi/3}\right) = V_\alpha - jV_\beta \tag{3.72}$$

$$V_\alpha = V_a; \quad V_\beta = -\frac{1}{\sqrt{3}}\left(V_c - V_b\right) = -\frac{1}{\sqrt{3}}\left(V_a + 2V_b\right); \quad V_a + V_b + V_c = 0 \tag{3.73}$$

$$i_\alpha = i_a; \quad i_\beta = -\frac{1}{\sqrt{3}}\left(i_a + 2i_b\right); \quad i_a + i_b + i_c = 0 \tag{3.74}$$

Only two voltages and two currents have to be measured:

TABLE 3.1.
Switching Pattern

SCR Conducting	$T_2T_5T_6$	$T_1T_4T_5$	$T_1T_4T_6$	$T_4T_3T_6$	$T_2T_3T_6$	$T_2T_3T_5$	$T_1T_3T_5$	$T_2T_4T_6$
$V(1)$	$V_1(0,0,1)$	$V_5(1,0,1)$	$V_4(1,0,0)$	$V_6(1,1,0)$	$V_2(0,1,0)$	$V_3(0,1,1)$	$V_0(1,1,1)$	$V_7(0,0,0)$
V_a	$-1/3\,V_0$	$-1/3\,V_0$	$+2/3\,V_0$	$+1/3\,V_0$	$-1/3\,V_0$	$-1/3\,V_0$	0	0
V_b	$-1/3\,V_0$	$-2/3\,V_0$	$-1/3\,V_0$	$+1/3\,V_0$	$+2/3\,V_0$	$+1/3\,V_0$	0	0
V_c	$+2/3\,V_0$	$+1/3\,V_0$	$-1/3\,V_0$	$-2/3\,V_0$	$-1/3\,V_0$	$+1/3\,V_0$	0	0

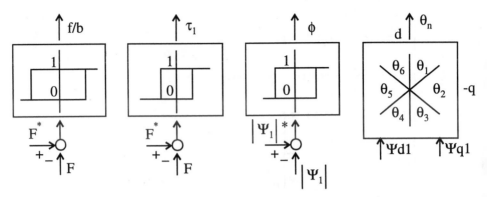

FIGURE 3.26
Table of adequate switchings (TAS).

TABLE 3.2
Flux Vector Location

sign λ_a	+	+	–	–	–	+
sign $(|\lambda_\alpha|-\sqrt{3}|\lambda_\beta|)$	+	–	–	+	–	–
sign λ_β	not relevant	+	+	not relevant	–	–
$\theta(i)$	$\theta(1)$	$\theta(2)$	$\theta(3)$	$\theta(4)$	$\theta(5)$	$\theta(6)$

$$\lambda_\alpha = \int (V_\alpha - R_1 i_\alpha)\, dt; \quad \lambda_\beta = \int (V_\beta - R_1\, i_\beta)\, dt \qquad (3.75)$$

Thus, the thrust (3.61) can be computed in real time also:

$$F_x = \frac{3\pi}{2\tau} (\lambda_\alpha i_\beta - \lambda_\beta i_\alpha) \qquad (3.76)$$

The sector $\theta(i)$, which is the location of the current flux vector, can be obtained by simple comparators (Table 3.2). If the number of poles is small (say, $2p = 2$ or $2p = 4$), the windings are not symmetric and the phase currents may not be symmetric either. To avoid unbalance in thrust and normal force distributions along the primary length, the i_α', i_β' are given by [11]

$$\begin{bmatrix} i'_\alpha \\ i'_\beta \end{bmatrix} = \frac{2}{3} \begin{bmatrix} 1 & -\dfrac{1}{2} & -\dfrac{1}{2} \\ 0 & -\dfrac{\sqrt{3}}{2} & \dfrac{\sqrt{3}}{2} \end{bmatrix} \begin{bmatrix} G_a i_a \\ G_b i_b \\ G_c i_c \end{bmatrix} \tag{3.77}$$

$$\lambda = C \sqrt{i'^2_\alpha + i'^2_\beta} \tag{3.78}$$

In (3.77), G_a, G_b, G_c are chosen by trial and error such that the i_1' vector,

$$i'_1 = i'_\alpha - j i'_\beta \tag{3.79}$$

is circular. This approximation is valid for small airgap variations. Otherwise the transformation (3.77) should be applied directly to the primary phase fluxes λ_a, λ_b, and λ_c. These fluxes are given by

$$\lambda_a = \int (V_a - R_1 i_a) \, dt$$

$$\lambda_b = \int (V_b - R_1 i_b) \, dt \tag{3.80}$$

$$\lambda_c = -(\lambda_a + \lambda_b)$$

3.8 DESIGN METHODOLOGY BY EXAMPLE

The design of an LIA is closely related to its potential application. Here we refer to a machine-tool table driven directly by a *flat single-sided LIA*. We illustrate the procedure by a numerical example.

Initial Data

Rated thrust, with natural cooling, $F_{xn} = 400$ N
Peak short-duration standstill thrust, $F_{xk} = 800$ N
Maximum travel length = 2 m
Rated speed $U_n = 2$ m/s
Power source: three-phase 220-V wye-connected, diode rectifier PWM voltage source
          inverter

Given these data, some additional practical information is needed to reduce the number of unknowns, including: mechanical or magnetic airgap $g = 1$ mm.

The airgap flux density for rated thrust may be chosen as $B_{gin} \approx 0.5$ to 0.6 T. Slot geometry is the same as for rotary induction motors: slot height/slot width less than 4 to 6 to allow for natural cooling and to keep the slot leakage low. The secondary ladder (or cage) winding in terms of number of slots per $2p$ poles of primary should follow the rules for rotary induction machines, to reduce the synchronous parasitic thrust components. The primary winding used here has two layers, full-pitch coils, and $(2p + 1)$ poles with half-filled end slots. The pole pitch τ depends on the performance criteria combination. Smaller τ means thinner back cores but lower power factor, though with shorter end connections of the coils the copper losses might not increase. Larger τ means better power factor but also longer end connections of coils and thicker back cores. As a start let us consider that the rated specific thrust with natural cooling is $F_x \approx 1$ N/cm^2. About 2 N/cm^2 can be obtained with forced air cooling and up to 2.5 to 3 N/cm^2 with liquid cooling.

Equivalent Circuit Parameters

The expressions for the parameters of the equivalent circuit of Fig. 3.18 for single-sided flat structures are as follows:

Primary resistance/phase,
$$R_1 = \frac{2\rho_{co}(l + l_{ec})J_{co}N_1^2}{N_1I_{1n}} \tag{3.81}$$

Primary leakage inductance,
$$L_{1\sigma} = \frac{2\mu_o}{pq}\left[(\lambda_{s1} + \lambda_{d1})l + \lambda_{ec}l_{ec}\right]N_1^2 \tag{3.82}$$

Slot-specific permeance,
$$\lambda_{s1} = \frac{h_{s1}(1 + 3p)}{12b_{s1}} + \frac{h_{sp}}{b_{sp}} \tag{3.83}$$

Airgap-leakage-specific permeance,
$$\lambda_{d1} = \frac{5(k_cg/b_{sp})}{5 + 4(k_cg/b_{sp})} \tag{3.84}$$

End-connection-specific permeance,
$$\lambda_{ec} = 0.3(3\beta - 1)q \tag{3.85}$$

Magnetizing inductance,
$$L_{1m} = \frac{6\mu_0 l\tau(k_{w1}N_1)^2}{\pi^2 k_c(1 + k_s)p_e g} \tag{3.86}$$

Secondary resistance, referred to the primary,

$$R_2' = 12p_{al} \frac{(k_{w1}N_1)^2}{N_{sa}} \left(\frac{l}{A_{s2}} + \frac{l_l}{A_l} \right) \tag{3.87}$$

Secondary leakage inductance, referred to the primary,

$$L_{2\sigma}' = 24\mu_0 l \left(\lambda_{s2} + \lambda_{d2} \right) \frac{(k_{w1}N_1)^2}{N_{sa}} \tag{3.88}$$

where

$$A_l = \frac{A_{s2}}{2\sin^2(\alpha_{es}/2)} \tag{3.89}$$

$$\alpha_{es} = \frac{\pi}{N_{sa}} (2p + 1) \tag{3.90}$$

$$l_l = \frac{(2p + 1)\tau}{N_{sa}} \tag{3.91}$$

$$\lambda_{s2} = \frac{h_{s2}}{3b_{s2}} + \frac{h_{ss}}{b_{ss}} \tag{3.92}$$

$$\lambda_{d2} = \frac{5k_c g/b_{ss}}{5 + 4k_c g/b_{ss}} \tag{3.93,}$$

Other symbols in (3.81) through (3.93) are as follows: N_{sa} is secondary slots per $(2p + 1)$ or $2p$ poles; p number of pole pairs; A_{s2} secondary active slot area; A_l secondary shorting bar cross section; l_l length of short bar per secondary slot; q number of slots per pole per phase; k_c Carter coefficient; g airgap; l_{ec} (0.01 + 1.5, coil span) = ρ_{co} resistivity of copper; J_{co} design current density; N_1 number of primary turns per phase, l stack width, k_{w1} primary winding factor; I_{1n} operating (full-load) current; b_{ss}, h_{ss}, b_{s2}, h_{s2}, etc., slot dimensions shown in Fig. 3.27; k_s saturation factor (< 1); β coil span/coil pitch; and τ pole pitch.

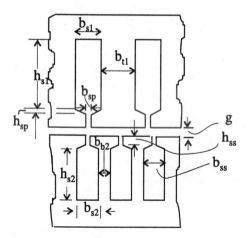

FIGURE 3.27
Slot geometry.

For aluminum sheet on iron secondary, the magnetic airgap becomes $g + d_{al}$, where d_{al} is the aluminum sheet thickness. Also $L_{2\sigma}'$ and, in the expression for R_2', $l_1 = 0$, $N_{sa}A_{s2} = 2pd_{al}$; ρ_{al} is augmented by a transverse edge effect coefficient $K_{fe} > 1$ [1-7], which accounts for return current paths in the secondary aluminum plate. The saturations factor k_s can be calculated as in rotary induction machines.

Basic Design Relationships

The airgap flux density under load B_{g1} with $L_{\sigma2}' \approx 0$ is given by [1]

$$B_{g1} = \frac{\mu_0 F_1}{g k_c (1 + k_s) \sqrt{1 + s^2 G_e^2}} \tag{3.94}$$

where

$$F_1 = \frac{3 \sqrt{2} \, N_1 K_{w1} I_1}{\pi p_e} \tag{3.95}$$

For the flat configuration the goodness factor G_e is

$$G_e = \frac{\omega L_{1m}}{R_2'} = \frac{\mu_0 \omega_1 \tau^2 h_{s2} (p + 1/2) \sigma_{al}}{2\pi^2 (p + 1/4) g k_c (1 + k_s) K_r} \tag{3.96}$$

and

$$K_r = 1 + \frac{L_l}{A_l} \frac{A_{s2}}{l} = 1.05 \text{ to } 1.25 \tag{3.97}$$

It can be demonstrated that the peak thrust per unit primary current (or mmf) is obtained for $sG_e = 1$. For the peak thrust, at standstill ($s = 1$) the airgap flux density given by (3.94) should not be above 0.7 T to avoid severe saturation. With $B_{g1k} = 0.7$ T, $s = 1$, and $G_e = 1$ from (3.94) we obtain

$$B_{g1k} = 0.7 = \mu_0 \frac{F_{1k}}{gk_c(1 + k_s) \sqrt{2}} \tag{3.98}$$

Considering for the time being $k_c \approx 1.25$ and $k_s = 0.4$, $g = 1$ mm, the peak primary mmf per pole F_{1k} is

$$F_{1k} = \frac{0.7 \times 10^{-3} \times 1.25 \times 1.4 \times 1.41}{1.256 \times 10^{-6}} = 1.375 \times 10^3 \text{ At/pole} \tag{3.99}$$

Also for $q = 2$ and $y/\tau = 5/6$ the primary winding factor k_{w1} is

$$k_{w1} = \frac{\sin \pi/6}{q \sin \pi/6q} \sin \frac{\pi}{2} \frac{y}{\tau} = \frac{0.5}{2 \times \sin \pi/12} \times \sin \frac{5}{6} \frac{\pi}{2} = 0.933 \tag{3.100}$$

The lower limit for the number of poles, to limit the phase current asymmetries, is $2p + 1 = 7$. Thus,

$$N_1 I_{1k} = \frac{F_{1k}\pi p_e}{3\sqrt{2} \, k_{w1}} = \frac{1.375 \times 10^3 \, \pi \, (3 + 0.25)}{3\sqrt{2} \times 0.933} = 3.55 \times 10^3 \text{ At/pole} \tag{3.101}$$

As a start let us use $l/\tau = 2.4$ in an attempt to reduce the coil end connection length. With a rated specific thrust $f_{xn} = 1 \text{ N/cm}^2$, the primary stack area A_p is:

$$A_p = (2p + 1)\tau l = (2p + 1)\tau^2 \frac{l}{\tau} = \frac{F_{xn}}{f_{xn}} = \frac{400\text{N}}{1\,\text{N/cm}^2} = 4 \times 10^{-2} \text{ m}^2$$

Therefore,

$$\tau = \sqrt{\frac{A_p \tau / l}{2p + 1}} = \sqrt{\frac{4 \times 10^{-2}}{7 \times 2.4}} = 4.88 \times 10^{-2} \text{ m}$$

For $q = 2$ it would mean a slot pitch $\tau_{s1} = \tau/6 = 8.14$ mm, which we believe is too small. So we come back to set $q = 1$, $y = \tau$, and thus $k_w = 1$ with a slot pitch $\tau_{s1} = \tau/3 = 16.22$ mm, which is reasonable. Now $N_1 I_{1k}$ can be decreased 0.933 times as $k_{w1} = 1$. For the time being we may leave it as is, for safety.

For the rated thrust we consider again $sG_e = 1$. We may choose the rated secondary frequency $f_2 = 5$ Hz to avoid severe noise and vibration at start (with $f_1 = f_2$) and to reduce the secondary slot cross section. For $sG_e = 1$ in (3.96) the only unknown is h_{s2} (the secondary active slot height). With $K_r = 1.2$, $k_c = 1.2$, and $k_{sn} = 0.3$ from (3.96) we obtain

$$h_{s2} = \frac{2\pi(3 + 0.25) \, 10^{-3} \times 1.2 \times 1.3 \times 1.2}{4\pi \times 10^{-7} \times 2 \times 5 \times 0.0488^2 \, (3 + 0.5) \, 3.2 \times 10^7} = 1.124 \times 10^{-2} \text{ m}$$

(3.102)

This is an acceptable value. It looks small but we have to consider that the travel (secondary) length is 2 m while the primary active length is $(2p + 1)\tau = 0.35$ m. Only one-sixth of secondary is always active, and consequently, in terms of both costs and overheating problems, the solution is acceptable. Now, as $sG_e = 1$ at both rated and peak thrust, and the thrust is proportional (in this case) to current squared, the rated phase ampere-turn $N_1 I_{1n}$ becomes

$$N_1 I_{1n} = N_1 I_{1k} \sqrt{\frac{F_{xn}}{F_{xk}}} = \frac{N_1 I_{1k}}{\sqrt{2}} = 2.334 \times 10^3 \text{ At/phase}$$

(3.103)

The primary active slot area A_{ps} can now be computed:

$$A_{ps} = \frac{N_1 I_{1k}}{pq J_{con} k_{fill}} \approx \frac{2.334 \times 10^3}{3 \times 1 \times 4 \sqrt{2} \, 10^6 \times 0.45} = 306.66 \times 10^{-6} \text{ m}^2$$

(3.104)

with $J_{con} = 4 \sqrt{2}$ A/mm^2 and k_{fill} = slot fill factor = 0.45. Now with $B_{g1k} = 0.7$ T and the primary slot pitch $\tau_s = 16.22$ mm, the slot width b_{s1} may be chosen as $b_{s1} = 9$ mm. Consequently, the primary slot active height h_{s1} is

$$h_{s1} = \frac{A_{ps}}{b_{s1}} = \frac{306.66}{9} = 43.7 \text{ mm} \tag{3.105}$$

The peak normal attraction force F_{nak} is

$$F_{nak} \approx \frac{B_{g1k}^2}{2\mu_0}(2p + 1)\tau^2\frac{l}{\tau} = \frac{0.7^2 \times 7 \times 0.0488^2 \times 2.4}{2 \times 1.256 \times 10^{-6}} = 7.804 \text{ kN} \tag{3.106}$$

For the rated thrust, the normal force is two times smaller. Still F_{na} is almost ten times higher than the thrust. Now we should check the rated thrust F_x as the 1 N/cm$_2$ thrust density is not guaranteed [1]. The rated thrust is given by

$$F_x = \frac{3I_1^2 l_{1m}}{(\tau/\pi)sG_e[(1/sG_e)^2 + 1]} \tag{3.107}$$

With $sG_e = 1$, (3.107) yields

$$F_{xn} = \frac{3\pi}{2\tau}I_{1n}^2 L_{1m} \tag{3.108}$$

But L_{1m} from (3.86) is

$$L_{1m} = \frac{6 \times 1.256 \times 10^{-6} \times 0.0488 \times 0.117}{\pi^2 \times 3.25 \times 10^{-3} \times 1.2 \times 1.4} \times N_1^2 = 8 \times 10^{-7} \times N_1^2 \tag{3.109}$$

Thus, (3.108) and (3.109) give

$$F_{xn} = \frac{3\pi}{2\tau}(N_1 I_{1k})^2 \times 8 \times 10^{-7} = \frac{3\pi(2.334 \times 10^3)^2 \times 8 \times 10^{-7}}{2 \times 0.0488} = 486\text{N} \tag{3.110}$$

So the rated thrust obtained is even greater than the predicted one. For safety the saturation factor k_s was kept as 0.4 and not 0.3 for rated current (thrust) conditions.

Number of Turns Per Phase N_1 and Performance

With $G_e = 1$, $f_1 = 5$ Hz at $s = 1$, the secondary resistance R_2' is

$$R_2' = \omega_1 L_{1m} = 2 \times \pi \times 5 \times 0.8 \times 10^{-6} N_1^2 = 2.512 \times 10^{-5} N_1^2 \quad (3.111)$$

Finally the primary resistance R_1 and leakage inductance are calculated as follows:

$$R_1 = \rho_c \frac{[2l + 2(0.01 + 1.5\tau)] \, j_{con} N_1^2}{N_1 I_{1n}}$$

$$= \frac{2.3 \times 10^{-8}[2 \times 0.117 + 2(0.01 + 1.5 \times 0.0488)] \times 4 \sqrt{2} \times 10^6 N_1^2}{2.51 \times 10^3}$$

$$= 2.07 \times 10^{-5} N_1^2 \quad (3.112)$$

$$\lambda_{s1} = \frac{34}{3 \times 9} + \frac{1.5}{3} = 1.76; \quad \lambda_{d1} = \frac{5 \times (1 \times 1.2/3)}{5 + 4(1 \times 1.2/3)} = 0.3 \quad (3.113)$$

$$\lambda_{ec} = 0.6 \quad (3.114)$$

$$l_{ec} = 0.01 + 1.5 \times 0.048 = 0.082 \text{ m} \quad (3.115)$$

$$L_{1c} = \frac{2\mu_0}{pq} [(\lambda_{s1} + \lambda_{d1})l + \lambda_{ec} l_{ec}] \, N_1^2 = \frac{2 \times 1.256 \times 10^{-6}}{(3 \times 1)}$$

$$\times [(1.76 + 0.3) \times 0.117 + 0.6 \times 0.082] N_1^2 = 0.243 \times 10^{-6} N_1^2 \quad (3.116)$$

We may now choose a number of slots in the active secondary N_{sa} greater than $N_{s1} = (2p + 1)m = 7 \times 3 = 21$ slots of the primary. Let us choose $N_{sa} = 30$ secondary slots per $(2p + 1)\tau$ length of secondary. The secondary slot pitch τ_{ss} then becomes

$$\tau_{ss} = \frac{(2p + 1)\tau}{N_{sa}} = \frac{7 \times 0.048}{30} = 1.12 \times 10^{-2} \text{ m} \quad (3.117)$$

The secondary slot width $b_{ss} = \tau_{ss}/2 = 5.6$ mm. As the secondary slot useful (active) height $h_{s2} = 14.4$ mm, the short-circuiting bar cross section A_L becomes

$$A_L = \frac{A_{s2}}{2 \sin 2 (\alpha_{es}/2)} = \frac{h_{s2} \times b_{s2}}{2 \sin^2 \dfrac{(2m + 1)\pi}{2N_{sa}}} = \frac{11.4 \times 5.7}{2 \times \sin^2 \dfrac{7\pi}{2 \times 30}} = 252.98 \text{ mm}^2$$

$$(3.118)$$

Thus, from (3.97), K_r becomes

$$K_r = 1 + \frac{11.38}{252.98} \times \frac{64.98}{117} = 1.02 \qquad (3.119)$$

As K_r was considered equal to 1.2, the height of the secondary slot may be reduced by the ratio 1.2/1.02 = 1.17. Alternatively, the primary frequency at start f_1 (or the rated secondary frequency f_{2n}) may be decreased to

$$f'_{2n} = \frac{f_{2n}}{1.17} = \frac{5}{1.17} = 4.27 \text{ Hz} \qquad (3.120)$$

Here we prefer this second alternative to keep the secondary losses low. The secondary leakage inductance $L_{2\sigma}$ is now computed from (3.88) after λ_{s2} and λ_{d2} are calculated from (3.92) and (3.93). Thus,

$$\lambda_{s2} = \frac{11.4}{3 \times 5.7} + \frac{1}{3} = 1.0$$

$$(3.121)$$

$$\lambda_{d2} = \lambda_{d1} = 0.3$$

and

$$L'_{2\sigma} = \frac{24 \times 1.256 \times 10^{-6} \times 0.117 \times (1.0 + 0.3)}{30} \times N_1^2$$

$$= 0.1528 \times 10^{-6} N_1^2 \qquad (3.122)$$

Up to this point we have neglected the effect of leakage inductance on the thrust. To compensate for it, we restore $f_2 = 5$ Hz = constant. The primary frequency f_{1n} for rated speed $u_n = 2$ m/s is thus

$$f_{1n} = \frac{u_n}{2\tau} + f_{2n} = \frac{2}{2 \times 0.048} + 5 = 25.833 \text{ Hz} \qquad (3.123)$$

We recall that in the expressions for the various parameters, the only unknown is N_1^2. We may now calculate N_1^2 for rated thrust conditions (F_{1n}, f_{2n}). The maximum rms phase voltage V_{1n} available from the inverter, for a 220-V ac wye-connected line voltage, is

$$V_{1n} = 220/\sqrt{3} = 127 \text{ V} \tag{3.124}$$

From the equivalent circuit of Fig. 3.20, with core loss neglected, we obtain

$$V_{1n} = I_{1n} \left| R_1 + j\omega_{1n}L_{1s} + \frac{j\omega_1 L_{1m}(R_2'\omega_1/\omega_2 + j\omega_2 + L_{2\sigma}')}{R_2'\omega_1/\omega_2 + j\omega_1(L_{1s} + L_{2\sigma}')} \right| \tag{3.125}$$

$$= N_1 \, (N_1 I_{1n}) \left| 2.225 \times 10^{-5} + j \, 2\pi \times 25.8 \times 0.243 \times 10^{-6} \right.$$

$$\left. + 2\pi \times 25.8 \left(j0.8 \times 10^{-6} \, \frac{(0.8 \times 10^{-6} + j0.1528 \times 10^{-6})}{0.8 \times 10^{-6} + j(0.8 \times 10^{-6} + 0.1528 \times 10^{-6})} \right) \right| \tag{3.126}$$

Finally with N_1, $I_{1n} = 2.334 \times 10^3$ A per phase and $V_{1n} = 127$ V,

$$N_1 = \frac{127}{2.334 \times 10^3 \times 10^{-5} \left| 2.25 + 5.4 + j10.537 \right|} \approx 420 \text{ turns/phase} \tag{3.127}$$

As there are 2 pq coils per phase (two layer windings), the number of turns per coil n_c is:

$$n_c = \frac{N_1}{2 \times 3} = \frac{420}{3 \times 2} = 70 \text{ turns/coil} \tag{3.128}$$

The rated phase current I_{1n} is

$$I_{1n} = \frac{N_1 I_{1n}}{N_1} = \frac{2.334 \times 10^3}{420} = 5.557 \text{ A} \tag{3.129}$$

The rated power factor is obtained from (3.127), the denominator of which has the total phase impedance. Thus,

$$\cos \phi_{1n} = \frac{7.625}{\sqrt{7.625^2 + 10.523^2}} = 0.5865 \qquad (3.130)$$

The rated efficiency η_{1n} from the same equation is

$$\eta_{1m} = \frac{5.4}{7.625} = 0.708 \qquad (3.131)$$

The power balance allows us to calculate the final value of thrust F_{xn} from

$$F_{xn} u_n = 3 V_{1n} I_{1n} \cos \phi_{1n} \, \eta_{1n} \qquad (3.132)$$

$$F_{xn} = \frac{3 \times 127 \times 5.557 \times 0.5865 \times 0.708}{2} = 440 \text{ N} \qquad (3.133)$$

So with $f_{2n} = 5$ Hz, $f_{1n} = 25.8$ Hz, and $V_1 = 127$ V the 400-N rated thrust is obtained. The power factor could have been better if the LIA force density had been lower, requiring more iron and more copper. The specific thrust f_{xwm} in terms of N/W is

$$f_{xwn} = \frac{F_{xw}}{F_{xn} u_n \left(\dfrac{1}{\eta} - 1 \right)} = \frac{1}{2 \left(\dfrac{1}{0.708} - 1 \right)} = 1.212 \text{ N/W} \qquad (3.134)$$

Or per volt-ampere, we have

$$f_{xvA} = \frac{F_{xn}}{3 V_{1n} I_{1n}} = \frac{440}{3 \times 127 \times 5.57} = 0.2075 \text{ N/VA} \qquad (3.135)$$

Primary Weight

The primary back-iron height h_{cp} is

$$h_{cp} \approx \frac{B_{g1}}{B_{cp}} \frac{2\tau}{\pi} = \frac{(0.7/\sqrt{2})}{1.2} \times \frac{2 \times 0.0488}{\pi} = 12.85 \times 10^{-3} \text{ m} \qquad (3.136)$$

For mechanical reasons we choose $h_{cp} = 15 \times 10^{-3}$ m. The primary weight is made essentially of copper and iron parts, G_{co} and G_{iron}. These are

$$G_{co} = 3N_1(2l + 0.02 + 3\tau)\frac{I_{1n}}{J_{con}}\gamma_{co} = 3 \times 420 \times 0.4 \times \frac{5.57}{5.64 \times 10^6} \times 8900$$

$$= 4.4196 \text{ kg} \tag{3.137}$$

where γ_{co} is the mass density of copper. Also,

$$G_{iron} = (2p + 1)\tau l\left(h_{cp} + \frac{1}{2}h_{sp}\right)\gamma_{iron}$$

$$= 0.35 \times 0.117 \times (15 + 37/2) \times 10^{-3} \times 7900 = 10.5139 \text{ kg} \tag{3.138}$$

So the total primary active material weight G_p is

$$G_p = G_{co} + G_{iron} = 4.4196 + 10.5139 = 14.9335 \text{ kg} \tag{3.139}$$

$$\frac{F_x}{G_p} = \frac{440}{14.9335} = 29.84 \text{ N/kg} \tag{3.140}$$

Considering that the peak short-duration thrust at standstill is $F_{xk} = 2F_{xn} = 880$ N, even with considerable frame weight, the primary alone can be accelarated from zero to 2 m/s at an average acceleration of $3a_g$ (where a_g is the acceleration due to gravity) over a distance of only 6.7×10^{-2} m. Of course the machine-tool table adds its mass to the primary, and the acceleration "on load" is considerably lower.

Note on Design Optimization

In the preceding sections an acceptable LIA has been designed. We now exercise optimization methods on it. Criteria for design optimization range from maximum efficiency power factor product to costs of active materials (primary plus secondary). More recently the net present worth or value (NPV) method has been applied to rotary machine design and may be extended to LIAs. The global costs of the LIA drive C_b is

$$C_b = C_{LIA} + C_{PE} + C_{loss} + C_{maint} \tag{3.141}$$

where C_{LIA} =   LIA costs
            C_{PE} =   power electronics costs
            C_{loss} =   loss capitalized costs
            C_{maint} =   maintenance costs

All of these cost components vary with time. So the money invested additionally in more copper and iron or power electronics to reduce the losses could have been invested at a certain interest rate and so on. To illustrate the NPV method let us consider the example of loss capitalized costs C_{loss}. Consider that in the first year S dollars have been spared on energy losses through a better design (with more active materials). Let us denote the interest rate by i and the yearly power loss cost increase by i_p and by i_E in effective interest rate per year. Then

$$i_E = \frac{1 + i}{1 + i_p} - 1 \tag{3.142}$$

Thus the NPV of losses for an n-year period is

$$\text{NPV} = S \, \frac{(1 + i_E)^n - 1}{i_E \, (1 + i_E)^n} \tag{3.143}$$

Now we can consider the influence of taxes and inflation of these savings:

$$\text{NPV}^e = \text{NPV}_e + \text{NPV}_D \tag{3.144}$$

where NPV_e is net present value of energy savings and NPV_D is net present value of depreciation on premium investment (straight line depreciation).

Therefore,

$$\text{NPV}_e = \text{NPV} \, (1 - T) \tag{3.145}$$

where T is the tax rage;

$$\text{NPV}_D = \frac{\text{NPV}_e}{n} \, \frac{[(1 + i)^n - 1)]}{i(1 + i)^n} \, T \tag{3.146}$$

Let us consider $S = 100$ dollars as first-year savings in energy losses over a period of $n = 19.4$ years with $i = 10\%$, $I_p = 8\%$, and $T = 40\%$. From (3.142) and (3.143) we obtain

$$NPV_\eta = 100 \times 16.18 = 1618 \tag{3.147}$$

And from (3.145) and (3.146) we have

$$NPV_e = 1618 \times 0.6 = 970.88 \tag{3.148}$$

$$NPV_D = \frac{970.8}{19.7} \times 8.43 \times 0.4 = 168.248 \tag{3.149}$$

Thus, (3.144) yields

$$NPV^e = 970.8 + 168.74 = 1139.8 \tag{3.150}$$

This value constitutes the maximum premium that can be expended to achieve $100 energy savings in the first year. In a similar way this rationale can be applied to initial costs or maintenance costs. Design changes also reduce the global costs C_b, keeping in mind that power electronic costs may exceed LIA costs.

In a similar manner a tubular LIA may be designed. Such an LIA is simple, robust, and inexpensive but in terms of thrust density, efficiency, and power factor is not particularly superior to other LIAs such as the one with permanent magnets to be treated in the next chapter.

REFERENCES

1. S. A. Nasar and I. Boldea, *Linear electric motors* (Prentice Hall, Englewood Cliffs, NJ, 1987).
2. E. R. Laithwaite, *Induction machines for special purposes* (George Newnes, London, 1966).
3. S. Yamamura, *Theory of linear induction motors* (Wiley-Interscience, New York, 1972).
4. S. A. Nasar and I. Boldea, *Linear motion electric machines* (Wiley-Interscience, New York, 1976).
5. P. K. Budig, *AC linear motors* (in German) (VEB Verlag, Berlin, 1978).
6. M. Poloujadoff, *The theory of linear induction machines* (book) (Clarendon Press, Oxford, 1980).
7. I. Boldea and S. A. Nasar, *Linear motion electromagnetic systems* (Wiley-Interscience, New York, 1985).
8. G. E. Dawson, A. R. Eastham, J. F. Gieras, R. Ong, and K. Ananthasivam, "Design of linear induction drives by finite element analysis and finite element techniques" *Rec. IEEE-IAS*, 1985 Annual Meeting, pp. 1762-1768.
9. S. Yamamura, *AC motors for high performance applications* (Marcel Dekker, New York, 1986).

10. I. Boldea and S. A. Nasar, *Vector control of a.c. drives* (CRC Press, Boca Raton, FL, 1992).
11. I. Takahashi and Y. Ide, "Decoupling control of thrust and attractive force of a LIA using space vector controlled invertor," *Rec. IEEE-IAS*, 1990 Annual Meeting.

LINEAR PERMANENT MAGNET SYNCHRONOUS ACTUATORS

Linear permanent magnet synchronous actuators (LPMSAs) have been proposed for applications in factory automation upgrading. Such actuators utilize high-energy product magnets and are characterized by high thrust density, low losses, small electrical time constant, and rapid response. The main drawbacks of LPMSAs are their high cost owing to the costs of the magnets and the presence of stray magnet fields, especially in single-sided configurations.

Thrust- or position-controlled flat LPMSAs may be used for travels up to about 3 m, with a constant airgap provided either mechanically or via a controlled suspension system. For travel lengths of less than 0.5 m tubular configurations may be preferred, since tubular structures make better use of materials, resulting in a compact actuator.

4.1 CONSTRUCTION DETAILS

LPMSAs may be flat single-sided or double-sided or may have a tubular structure. In either of these configurations, the primary may constitute the mover. On the other hand, the actuator may have moving magnets. The core of the primary of a flat LPMSA is made of longitudinal laminations with uniformly distributed slots, which house the windings. Often, the windings are three-phase distributed windings having one to four slots/pole per phase (q = 1, 2, 3, or 4). These windings are similar to those of induction-type actuators presented in Chapter 3. Because the windings are located in open slots, the effective airgap becomes considerably greater than the actual airgap.

In tubular structures the laminations of the primary core may be either longitudinal or disk-shaped as in linear induction actuators. Stacking of the disk laminations increases the effective airgap. The core of the secondary of a tubular LPMSA is generally made of solid magnetic steel. Rare-earth permanent magnets (PMs) in LPMSAs are capable of producing adequate flux densities in about 1-mm effective airgaps. Because of ease of assembly, tubular PMLSAs with disk-shaped laminations are often preferred to flat ones. In order to reduce eddy current losses the disk laminations may be radially slitted.

Linear bearings support and guide the motion of the mover and also maintain the desired airgap. Electromagnetic-controlled suspensions or air bearings are used to support the mover in special cases.

4.2   FIELDS IN LPMSAs

High-energy product rare-earth PMs are almost invariably used in LPMSAs to obtain large thrust/weight and thrust/power input ratios. Such magnets are characterized by large remnant flux densities B_r and large coercive forces H_c, as shown in Fig. 4.1. Their recoil permeabilities μ_{re} are in the range 1.02 to 1.18 μ_0. A typical high-energy product PM has a linear demagnetization characteristic of the form

$$B_m \approx B_r + \mu_{re} H_m \tag{4.1}$$

Consequently, such a PM can be replaced by an equivalent mmf, M_{PM}, given by

$$M_{PM} = H_c h_m \tag{4.2}$$

where h_m is magnetic thickness and H_c is coercive force. This equivalent (fictitious) mmf is located in air with $\mu_{re} \approx \mu_0$, and may be in the form of a current sheet. Such an approximation is often used with both analytical and finite-element methods (FEM) to obtain field distributions in various regions of LEAs [3]. It is to be noted that FEM is used only for design refinements, and not for a preliminary design of an LPMSA. In contrast to FEM, analytical methods yield quick insight into the various phenomena while requiring much less computation time.

The primary of a three-phase LPMSA is constructed as coils in slots and may be supplied by currents 120° out of phase from each other with two phases conducting at one time; or the primary may be supplied with a three-phase sinusoidal excitation. In both cases a PWM voltage source invertor is used. In the following, first we consider a flat LPMSA supplied with a three-phase sinusoidal current. The primary winding may have two, three, or four slots/pole per phase.

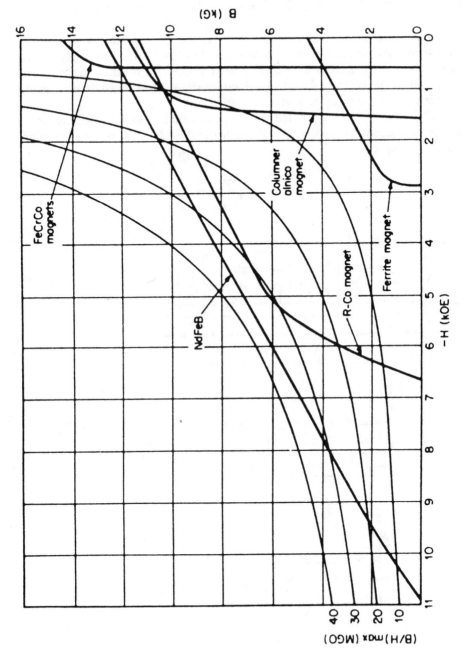

FIGURE 4.1
BH-characteristics of some rare-earth magnets.

4.3 SINUSOIDAL CURRENT MODEL FIELD DISTRIBUTION

For a single-sided flat LPMSA, we first use an analytical approach to find the fields and compare the results with those obtained by FEM [4,5]. Then we apply the *dq*-model to assess the actuator's steady-state performance. We calculate the thrust and normal forces and compare the calculated results with those obtained experimentally. The major assumptions in the analysis are as follows:

(i) The slotted structure of the primary is replaced by a smooth structure having anisotropic magnetic properties, with the permeabilities along the *x*- and *y*-directions (Fig. 4.2) given by

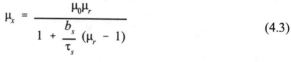

$$\mu_x = \frac{\mu_0 \mu_r}{1 + \dfrac{b_s}{\tau_s}(\mu_r - 1)} \tag{4.3}$$

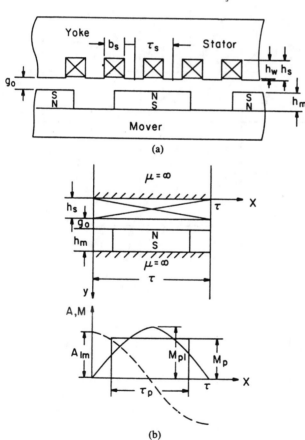

(a)

(b)

FIGURE 4.2
Single-sided LPMSA: (a) actual structure; (b) model for mathematical analysis.

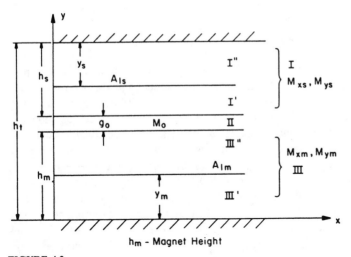

FIGURE 4.3
LPMSA model for analysis.

and

$$\mu_y = \mu_0 \left[\frac{b_s}{\tau_s} + \mu_r \left(1 - \frac{b_s}{\tau_s} \right) \right] \tag{4.4}$$

where μ_r is relative permeability, b_s is slot width (or opening), and τ_s is slot pitch.

(ii) Regions containing magnets are considered isotropic, with $\mu_x = \mu_y \approx \mu_0$.

(iii) Regions containing iron are assumed to be infinitely permeable.

(iv) Only the fundamental component of the primary mmf is included in the analysis.

(v) The fundamental component M_{1PM} of the PM mmf given by (4.2) is considered, where M_{1PM} is given by

$$M_{1PM} = \frac{4}{\pi} M_{PM} \sin \frac{\pi \tau_p}{2\tau} \sin \frac{\pi}{\tau} x \tag{4.5}$$

The corresponding equivalent current sheet becomes

$$A_{1m} = \frac{4}{\tau} M_{PM} \sin \frac{\pi \tau_p}{2\tau} \cos \frac{\pi}{\tau} x \tag{4.6}$$

where τ_p and τ are defined in Fig. 4.2.

(vi) The fundamental component of the armature reaction current sheet A_{1s} is expressed as

$$A_{1s} = \frac{3\sqrt{2}}{p\tau} K_{w1} N_1 I_1 \cos\left(\frac{\pi}{\tau}x + \gamma_0\right) \tag{4.7}$$

where K_{w1} is the primary winding factor, $N_1 I_1$ primary mmf, p number of poles, τ pole pitch, and γ_0 displacement between the primary and secondary equivalent current sheet.

The locations of the current sheets given by (4.6) and (4.7) are shown in Fig. 4.3. Consequently, we have three different regions in the equivalent model, as shown in Fig. 4.3. With $\mu_x \neq \mu_y$ in regions I/ and I//, an application of Maxwell's equations yields analytical solutions to the fields [4]. It is found that the locations of equivalent current sheets do not have any notable effects on the flux densities. For $b_s/\tau = 0.53$, $h_s = 35$ mm, $\tau = 114.3$ mm, $\tau_p/\tau = 0.67$, $\mu_r = 400$, $g = 2$ mm, $h_m = 6.35$ mm, $H_c = 0.7$ MA/m, $B_r = 0.86$ T, and primary stack width $l_A = 101.6$ mm, the normal airgap flux density calculated analytically, calculated by FEM, and obtained experimentally on a laboratory model are shown in Figs. 4.4 through 4.6. These illustrations of the various flux densities are for $\gamma = 90°$. The flux density pulsations owing to slot openings become evident in the results obtained by FEM and tests, although the fundamental components of the result obtained by FEM are not much different from the analytical results. Consequently, some confidence in the analytical approach is established, and we use it in the following.

The fundamental of the no-load induced voltage is

$$E_1 = \frac{\sqrt{2}}{\pi} \omega_1 K_{w1} N_1 \tau l_A (B_{yPM1})_{av} \tag{4.8}$$

where $(B_{yPM1})_{av}$ is the fundamental of the PM normal airgap flux density averaged over the primary slot height. Similarly, if I_1 is the primary current, the magnetizing inductance L_{am} can be expressed as

$$L_{am} = \frac{\sqrt{2}}{\pi I_1} K_{w1} N_1 \tau l_A (B_{ya1})_{av} \tag{4.9}$$

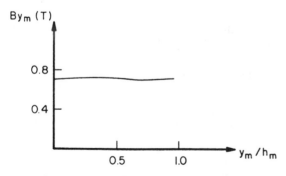

FIGURE 4.4
Influence of location of current sheets on PM fields.

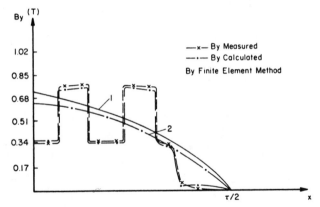

FIGURE 4.5
Normal flux density due to PM; fundamental obtained (1) by analysis and (2) by FEM.

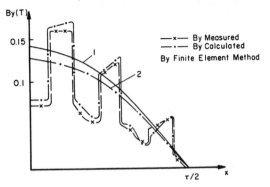

FIGURE 4.6
Airgap flux density due to the armature mmf; fundamental obtained (1) by analysis and (2) by FEM.

where $(B_{ya1})_{av}$ is the fundamental of the armature reaction normal flux density averaged over the primary slot height.

The leakage inductance $L_{1\sigma}$ has essentially two components. These are slot- and end-connection leakages, and can be determined by the method used for linear induction actuators. The same is true for the primary resistance R_1. The synchronous inductance L_s of an LPMSA is then given by

$$L_s = L_{1\sigma} + L_{am} \tag{4.10}$$

4.4 THE dq-MODEL AND FORCES

Because we are considering only the fundamental, we may use the dq-model for the actuator. Notice that the field mmf is constant, representing the magnets. Also, there is no damper winding on the secondary [5]. The dq-model equations are (with $p = d/dt$)

$$v_d = R_s i_d + p\lambda_d - \omega_r \lambda_q \tag{4.11}$$

$$v_q = R_s i_q + p\lambda_q + \omega_r \lambda_d \tag{4.12}$$

where

$$\lambda_d = L_s i_d + \lambda_{PM1} \tag{4.13}$$

$$\lambda_q = L_s i_q \tag{4.14}$$

and $\omega_r = \pi u / \tau$, u being the linear velocity.

The total flux from the PM is obtained from (4.8) such that

$$\lambda_{PM} = \frac{\sqrt{2} E_1}{\omega_r} = \frac{2}{\pi} K_{w1} N_1 \tau l_A (B_{yPM1})_{av} \tag{4.15}$$

The dq-transformation for voltages, currents, and fluxes is

$$\begin{bmatrix} i_d \\ i_q \end{bmatrix} = \frac{2}{3} \begin{bmatrix} \cos(-\theta_r) & \cos\left(-\theta_r + \dfrac{2\pi}{3}\right) & \cos\left(-\theta_r - \dfrac{2\pi}{3}\right) \\ \sin(-\theta_r) & \sin\left(-\theta_r + \dfrac{2\pi}{3}\right) & \sin\left(-\theta_r - \dfrac{2\pi}{3}\right) \end{bmatrix} \begin{bmatrix} i_a \\ i_b \\ i_c \end{bmatrix} \tag{4.16}$$

In (4.16) $\theta_r = \int \omega_r \, dt + \theta_{10}$ for the PM mover and $\theta_r = -\int \omega_r \, dt + \theta_{r0}$ for the primary (which moves against its own traveling field).

The developed electromagnetic power is given by

$$P_e = F_x u_s = F_x 2\tau f_1 = \frac{3}{2} \omega_1 (\lambda_d i_q - \lambda_q i_d) \tag{4.17}$$

The thrust F_x is expressed as

$$F_x = \frac{3\pi}{2\tau} \lambda_{PM} i_q \tag{4.18}$$

The normal force can be obtained from the derivative of the magnetic stored energy. Thus,

$$F_{na} = \frac{3}{2}\left(i_d \frac{\partial \lambda_d}{\partial g} + i_q \frac{\partial \lambda_q}{\partial g}\right) \tag{4.19}$$

In the actuator under consideration the d- and q-axis inductances are the same, and from (4.18) it follows that only i_q contributes to the thrust. Therefore, i_q = constant control via power electronics should be targeted.

We now combine (4.11) and (4.12) to obtain the space-phasor equation as follows:

$$v_1 = R_1 i_1 + p\lambda_s + j\omega_r \lambda_s \tag{4.20}$$

where

$$v_1 = v_d + j v_q; \quad i_1 = i_d + j i_q; \quad \lambda_s = \lambda_d + j\lambda_q \tag{4.21}$$

and $\omega_r = \pi u_s/\tau$. Under steady state in synchronous coordinates (with the axes fixed to the PM), $p x_s = 0$. Thus, the phasor diagram of Fig. 4.7 is obtained. Notice that this diagram does not have time-dependent quantities. Rather, it has phasors shifted by space angles, with the two orthogonal axes moving at the synchronous speed u_s or rotating at the speed ω_r.

With $i_d = 0$, the thrust varies with i_q, which, when controlled, yields the desired mechanical characteristics. Also, the normal force increases with the thrust, since the

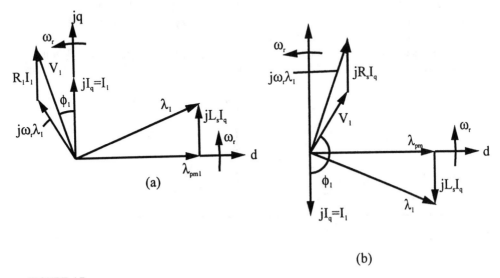

FIGURE 4.7
Phasor diagram for LPMSA: (a) motoring; (b) generating.

primary flux λ_1 (and, implicitly, the airgap flux $|\lambda_{1g}| = |\lambda_1 - L_{1\sigma}i_1|$ increases). The i_q-dependent characteristics are limited by the maximum voltage available from the inverter with a constant flux from the PM.

In order to obtain the thrust-speed characteristic we refer to the space-phasor diagram (Fig. 4.7), from which

$$V_1^2 = (\omega_r L_q i_q)^2 + (R_s i_q + \omega_r \lambda_{PM})^2 \tag{4.22}$$

Combining (4.20) and (4.22) yields, with $\omega_r = \pi u_s / \tau$,

$$V_1^2 = \left(\frac{2L_q u_s F_x}{3\lambda_{PM}}\right)^2 + \left(\frac{2R_s \tau F_x}{3\pi\lambda_{PM}} + \frac{\pi}{\tau} u_s \lambda_{PM}\right)^2 \tag{4.23}$$

For $i_d = 0$ and with a given voltage V_1, the thrust-speed characteristic is similar to that of a dc series motor. The ideal no-load speed is given by (Fig. 4.8)

$$(u_{so})_{F_x=0} = \frac{V_1 \tau}{\pi \lambda_{PM}} \tag{4.24}$$

As the thrust increases, saturation of the low-speed region limits the flux and almost a linear characteristic is obtained. To vary the thrust, the voltage is varied, while the frequency varies implicitly to maintain $i_d = 0$ and to keep i_q in phase with the motion of the secondary. In the generating mode, $i_q < 0$ and $F_x < 0$ and the braking force is slightly greater than that for motoring at a given speed.

Thrust and normal forces as a function of γ_0 are shown in Figs. 4.8 and 4.9 respectively. Notice the large value of the normal force compared with the thrust. In

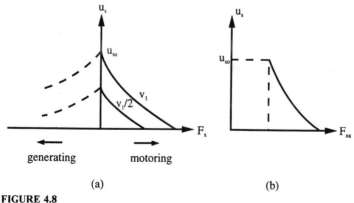

(a)   (b)

FIGURE 4.8
(a) Thrust vs. speed; (b) normal force vs. speed.

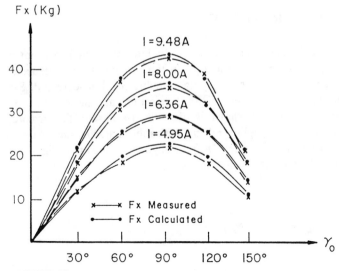

FIGURE 4.9
Thrust vs. γ_0.

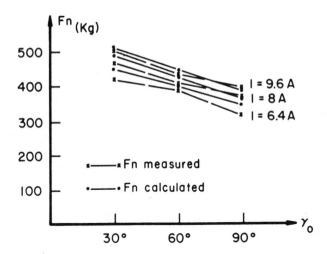

FIGURE 4.10
Normal force vs. γ_0.

Fig. 4.10, the current is kept constant with i_d decreasing from I_{dc} at $\gamma_0 = 0$ to zero at $\gamma_0 = \pi/2$. This fact explains the decrease of the normal force with γ_0 for a constant "total" current.

There are several other approaches to the analysis to PMLSAs, such as using the concept of magnetic charges [6] and a combination of FEM and harmonic decomposition techniques [7]. Finally, a comprehensive FEM analysis of fields and forces in a single-sided flat LPMSA having a two-pole mover is given in Reference [8].

4.5 RECTANGULAR CURRENT MODE

In the rectangular current mode, the phase currents ideally have the waveforms shown in Fig. 4.11, where two phases conduct at any given time, assuming instantaneous commutation. We consider such currents supplying a double-sided air-core PMLSA shown in Fig. 4.12. Air core is used to reduce thrust pulsations and the mass of the primary. The model used for analysis is shown in Fig. 4.13. Because of the air core, rare-earth high-energy magnets having a linear demagnetization characteristic are used. With the magnetization vector of the magnet, $M(x)$ rectangular, as shown in Fig. 4.13, we may decompose it into harmonics [9] as follows:

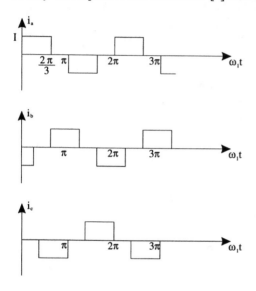

FIGURE 4.11
Ideal rectangular phase currents.

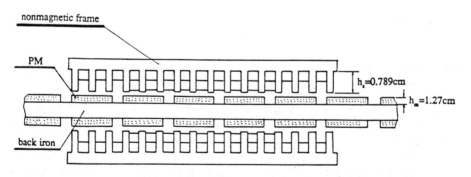

FIGURE 4.12
Double-sided air-core LPMSA.

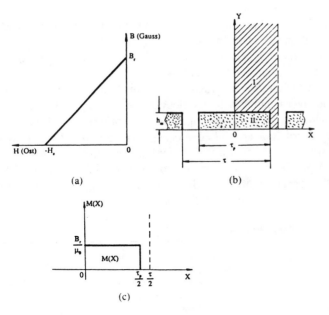

FIGURE 4.13
Model for mathematical analysis.

$$M(x) = \frac{4B_r}{\pi\mu_0} \sum_{m=1}^{\infty} \frac{1}{2m-1} \sin(2m-1)\pi\frac{\tau_p}{\tau} \cos(2m-1)\frac{\pi x}{\tau} \qquad (4.25)$$

The magnetic scalar potential satisfies Laplace's equation; it yields the flux densities in various regions of the model of [9]. For region I we have

$$B_{xl}(x,y) = \frac{4B_r}{\pi} \sum_{m=1}^{\infty} \frac{1}{2m-1} \sin(2m-1)\frac{\pi}{2\tau}\tau_p \sinh(2m-1)\frac{\pi}{\tau}h_m$$

$$\times \exp\left[-(2m-1)\frac{\pi}{\tau}y\right]\sin(2m-1)\frac{\pi x}{\tau} \qquad (4.26)$$

$$B_{yl}(x,y) = \frac{4B_r}{\pi} \sum_{m=1}^{\infty} \frac{1}{2m-1} \sin(2m-1)\frac{\pi}{2\tau}\tau_p \sinh(2m-1)\frac{\pi}{\tau}h_m$$

$$\times \exp\left[-(2m-1)\frac{\pi}{\tau}y\right]\cos(2m-1)\frac{\pi x}{\tau} \qquad (4.27)$$

From these expressions, it follows that the flux densities are dependent on B_r, τ_p/τ, and h_m/τ, and that they decay exponentially along OY. The fundamental decays by a factor e (= 2.718) at $y = \tau/\pi$. We may conclude that the primary winding should not be placed at a slot depth greater than τ/π from the magnet surface. In other words, the maximum depth of a slot $(h_s)_{max}$ is constrained by $(h_s)_{max} < (\tau/\pi - g)$. For an LPMSA, the design data are summarized in Table 4.1.

In order to determine the thrust and normal force of this PMLSA we assume that it has a full-pitch double-layer winding with one slot/pole per phase ($q = 1$) with only two phases (A^+B^-) conducting, as shown in Fig. 4.14. The primary, mmf, and PM average flux density, B_g, distribution are shifted in phase from 120° to 60° with phases A^+B^- conducting. The average phase shift is 90°. The primary has five poles with half-filled end slots. Thrust is produced by the interaction of the normal flux density of the magnet, $B_{ylav}(x,t)$, and the primary mmf. Similarly, interaction of $B_{xlav}(x,t)$ and primary mmf leads to the normal force $F_n(t)$. Phases A^+B^- begin to conduct simultaneously at

TABLE 4.1 Design Data for an LPMSA Prototype

PM type: samarium cobalt	Pole pitch $\tau = 1.15$ in.
$B_r = 9000$ G	Airgap $g = 0.04$ in.
$H_c = 8700$ Oe	Slot height $h_s = 0.35$ in.
Magnet height $h_m = 0.5$ in.	Slot pitch $\tau_s = \tau/3$
Magnet width $\tau_p = 1.0$ in.	Slot width $W_s = 0.7\,\tau_s$
	Tooth width $W_t = \tau_s - W_s$

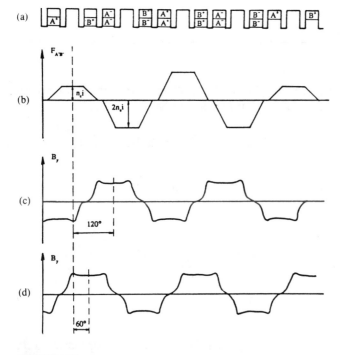

FIGURE 4.14
Relative position of PM and stator mmf.

$\omega_r t = 0$ and stop conducting at $\omega_r t = \pi/3$. The PM flux density is shifted by 120° with respect to the primary mmf at $t = 0$ and by 60° at $\omega_r t = \pi/3$. With this in mind, the ultimate expressions for the forces become [9]

$$F_x(t) = \frac{64 n_c I l}{w_s} \sum_{m=1}^{\infty} C_m \cos(2m - 1)\frac{\pi}{6} \left[\cos(2m - 1)\left(\frac{\pi}{\tau}ut - \frac{\pi}{6}\right)\right] \tag{4.28}$$

$$F_y(t) = \frac{64 n_c I l}{w_s} \sum_{m=1}^{\infty} C_m \sin(2m - 1)\frac{\pi}{6} \left[\cos(2m - 1)\left(\frac{\pi ut}{\tau} + \frac{\pi}{6}\right)\right] \tag{4.29}$$

where

$$C_m = \frac{4 B_r \tau^2}{\pi^3 h_s (2m - 1)^3} \sin(2m - 1)\frac{\pi \tau_p}{2\tau} \sinh(2m - 1)\frac{\pi}{2}h_m \exp[-(2m - 1)\frac{\pi}{\tau}h_m]$$

$$\times \left[1 - \exp[-(2m - 1)(h_s + g)\frac{\pi}{\tau}] \sin(2m - 1)\frac{\pi h_s}{2\tau}\right] \tag{4.30}$$

Also, l is stack width, w_s slot width, h_s slot depth, u mover speed, h_m magnet thickness, and g airgap. Plots of forces, as given by (4.28) and (4.29), are shown in Fig. 4.15. Notice the force pulsations over a 60° period when phases A^+B^- are conducting. To obtain fewer pulsations in the forces, nonrectangular currents are required. Because of the air core, the normal force is small. An increase in thrust can be obtained at the expense of increased copper losses by increasing the primary current.

4.6 DYNAMICS AND CONTROL ASPECTS

Power electronics control is invariably required in the operation of LPMSAs. Two basic strategies are rectangular and sinusoidal current control. In either case, the dynamics of an LPMSA can be formulated in phase coordinates, since the primary self-inductances are independent of the mover position. If L_s is the synchronous inductance for a three-phase primary we have

$$v_a = R_1 i_a + L_s p i_a + e_a \tag{4.31}$$

$$v_b = R_1 i_b + L_s p i_b + e_b \tag{4.32}$$

$$v_c = R_1 i_c + L_s p i_c + e_c \tag{4.33}$$

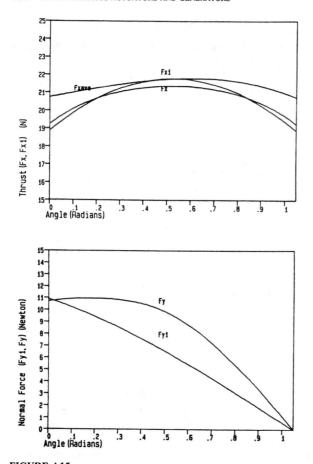

FIGURE 4.15
Thrust and normal force pulsation with rectangular current.

where

$$e_{a,b,c} = \sum_{n=1}^{\infty} C_n u \cos\left[\frac{\pi u t}{\tau} + \gamma_0 - (i-1)\frac{2\pi}{3}\right], \quad i = 0, \frac{2\pi}{3}, -\frac{2\pi}{3} \quad (4.34)$$

$$i_a + i_b + i_c = 0 \quad (4.35)$$

$p = d/dt$, u is mover velocity, and R_1 is primary resistance per phase. The thrust can be expressed in terms of electromagnetic power such that

$$F_x = \frac{1}{2u}\left(e_a i_a + e_b i_b + e_c i_c\right) \quad (4.36)$$

The mechanical equations are

$$M\dot{u} = F_x - F_{load} \qquad (4.37)$$

$$\dot{x} = u \qquad (4.38)$$

Equations (4.31) through (4.38) yield a state-space system with i_a, i_b, i_c, u, and x as state variables.

Reconsidering (4.31) through (4.33), L_s for $q \geq 2$ can be expressed as

$$L_s = L_{aa} - L_{ab} \qquad (4.39)$$

where L_{aa} is the self-inductance of phase a and L_{ab} is the mutual inductance between phases a and b of the primary. For slots/pole per phase, $q = 1$, the winding distribution is nonsinusoidal and the inductances L_{aa} and L_{ab} are given by [1]

$$L_{aa} = L_{1\sigma} + \frac{\mu_0 N_1^2 \tau l}{2p(g + h_m)} \qquad (4.40)$$

$$L_{ab} = - (L_{aa} - L_{1\sigma})/3 \qquad (4.41)$$

where p is the number of poles, $L_{1\sigma}$ primary leakage inductance, and h_m magnet thickness. Finally, for a sinusoidally distributed winding, we have

$$L_{aa} = L_{1\sigma} + \frac{4\mu_0 K_{w1}^2 N_1^2 \tau l}{\pi^2 p(g + h_m)} \qquad (4.42)$$

$$L_{ab} = - \left(L_{aa} - L_{1\sigma}\right)/2 \qquad (4.43)$$

With the parameters known, $e_{a,b,c}(t)$ can be determined by a numerical method, and thereby the actuator dynamics obtained in phase coordinates.

Rectangular Current Control

Rectangular current control uses a six-element position sensor while the amplitude of the current is kept constant by adequate chopping. Such a control (Fig. 4.16) has been

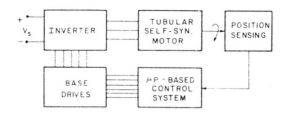

FIGURE 4.16
Block diagram of a rectangular current controller.

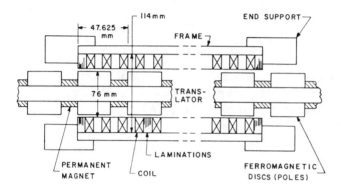

FIGURE 4.17
Cross section of a tubular LPMSA.

TABLE 4.2 Design data of the actuator of Fig. 4.18.

Item	Value
Bore diameter	76 mm
Pole pitch	47.625 mm
Airgap length	1 mm
Outside diameter	114 mm
No. of stator poles	8
No. of translator poles	11
Turns per coil	320
Wire size	AWG 29
Resistance per coil (20°C)	16.7 Ω
Inductance per coil (translator removed)	76.6 mH
Total no. of coils	24
Connection	Series wye

applied to a tubular LPMSA [10]. The actuator is shown in Fig. 4.17. The data pertaining to the LPMSA are given in Table 4.2. In essence, a six-element Hall-type sensor produces the signals to switch, in proper sequence, the two conducting phases. The output of the rectangular current is controlled by a position and speed cascaded controller whose output is the reference current amplitude I^*. Alternatively, chopping the voltage to maintain I^* over two consecutive commu-tations in each phase may be

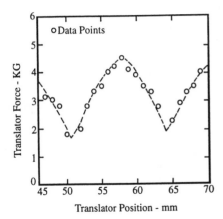

FIGURE 4.18
Thrust pulsation with position.

done in the upper- or lower-leg SCRs of the inverter. Thus, the switching frequency of the SCR is reduced, as the turn-on and turn-off signals for phase commutation are required only once in 60° (or a distance of $\tau/3$). Owing to the use of ferrite magnets in the LPMSA of Table 4.2, the thrust pulsates as shown in Fig. 4.18, as the d- and q-axis magnetizing inductances are unequal. Further details regarding the oscillatory behavior of this actuator are available in Reference [10].

Sinusoidal Current Control

As stated earlier, with the use of high-energy magnets ($L_s = L_d = L_q$) sinusoidal current control is simplified in that i_q vector current control is performed. Such a control system is shown in Fig. 4.19. Position and speed controllers provide the reference thrust F_x^*, which is proportional to i_q^* with the constant $2\tau/(3\pi\lambda_{PM})$. The linear position sensor also generates the Park transformation angle θ_r, which is multiplied by -1 if the primary is the mover, moving against the direction of the traveling field. Present-day microcontrollers are used to perform all control functions.

The precise position sensor required for a sinusoidal current control may be replaced by a position observer based on the two measured voltages, currents v_a, v_b, i_a, and i_b, known PM flux λ_{PM}, synchronous inductance L_s, and the primary resistance R_1.

As an example, we present a magnetically suspended vehicle [11] as an LPMSA (Fig. 4.20). The primary is 1 m long along the track. There are four two-pole PM translators, weighing 26.2 kg, and having control coils for controlled suspension. Assuming a constant 3.2-mm airgap, the fluxes produced by the controlled magnets may be considered constant. For simplicity, in the following we investigate the propulsion force developed by one of the four translators. This LPMSA has a speed control system, which may be either open-loop or closed-loop. Open-loop control is used at very low speeds, where a precise speed feedback is not feasible. In this case we have

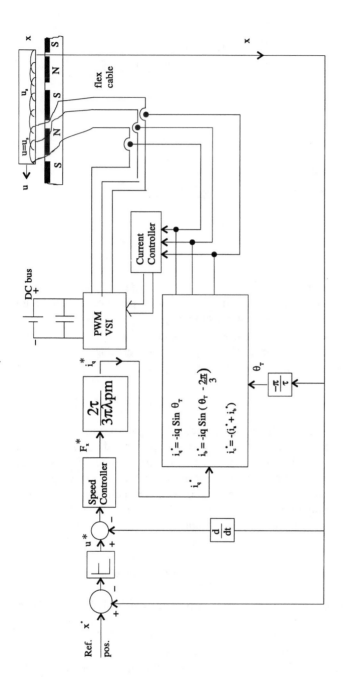

FIGURE 4.19
Indirect i_q-current vector control system.

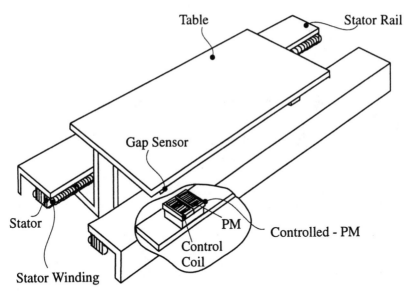

FIGURE 4.20
LPMSA Maglev carrier [11].

$$M\dot{u}_x = 4F_x \tag{4.44}$$

$$F_x = k_f I_q \tag{4.45}$$

where k_f is a proportionality constant. Based on the reference speed profile $u^*(t)$, $F_x^*(t)$ and $I_q^*(t)$ are, respectively, obtained from (4.44) and (4.45). The reference phase currents are

$$i_a^* = -I_q^* \sin \theta_r^*$$

$$i_b^* = -I_q^* \sin\left(\theta_r^* - \frac{2\pi}{3}\right) \tag{4.46}$$

$$i_c^* = -(i_a^* + i_b^*)$$

with

$$\theta_r^* = \frac{\pi}{\tau} \int u_x^* \, dt \tag{4.47}$$

The reference phase voltages are obtained from (4.31) through (4.33), where the currents and their derivatives are given by (4.46) and (4.47), and $\gamma_0 = \pi/2$. Such an open-loop control system has been demonstrated to perform well [11], as shown in

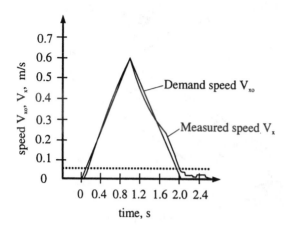

FIGURE 4.21
Open-loop speed dynamics.

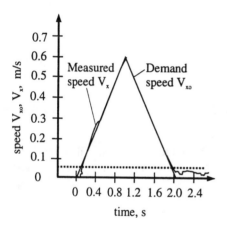

FIGURE 4.22
Closed-loop speed dynamics.

Fig. 4.21, especially during acceleration. In order to improve the response, a speed sensor u_x is used to generate the amplitude of the primary voltage V_1 such that

$$V_1 = G_1(u_x^* - u_x) + G_2 \int (u_x^* - u_x) \, dt + V_{10} \qquad (4.48)$$

During acceleration G_1 and G_2 are positive, and during deceleration they are negative; V_{10} is the open-loop precalculated voltage, which is neglected at speeds higher than 0.1 m/s. With such a control the speed profile is better, as shown in Fig. 4.22.

4.7 DESIGN METHODOLOGY

We illustrate the design procedure by designing a tubular LPMSA having a PM mover required to fit within a pipe of 50 mm internal diameter. The actuator uses a rectangular current control via a PWM voltage-source inverter. Other specifications are as follows

Basic Constraints

Maximum external (stator) diameter = 0.047 m
Stroke length = 0.61 m
Average linear speed = 0.216 m/s
Mode of operation: oscillatory
Environment: corrosive
Thrust = 114 N for continuous duty or 227 N for 38% duty cycle

Magnetic Circuit Design

Given the small external diameter of the stator, a tubular structure is the most appropriate choice. In order to utilize the iron and copper volume maximally, ring-shaped stator coils are placed in open slots. Consequently, no end connections of the windings exist and almost the entire amount of copper (in slots) is useful. The magnetic circuit geometry, along with the locations of the coils, is shown in Fig. 4.23. Ring-shaped PMs are mounted on the mover. As the mover, with these magnets, moves, motional voltages are induced in the three-phase stator windings and in the stator core, and hysteresis and eddy current losses are produced in the core. The core losses depend on the core flux density, which should be fairly high to reduce the volume, and on the frequency. The frequency f_1 is dependent on the synchronous speed u_s and stator winding pole pitch τ, such that

$$f_1 = u_s/2\tau \tag{4.49}$$

Reducing the pole pitch to the mechanically feasible minimum value of $\tau = 3$ cm is required to reduce the back-iron core height both in the secondary and in the primary (or stator), as the external diameter of the stator is much too small. From (4.49) it follows that a small pole pitch will increase the frequency f_1 and thereby the core losses will increase. The operating frequency, from (4.49), is

$$f_1 = \frac{0.2159}{2(0.03)} = 3.5983 = 3.6 \text{ Hz} \tag{4.50}$$

And the travel time t_t over one stroke length at a constant speed is

$$t_t = \frac{\text{stroke length}}{\text{speed}} = \frac{0.61}{0.2159} = 2.882 \text{ s} \tag{4.51}$$

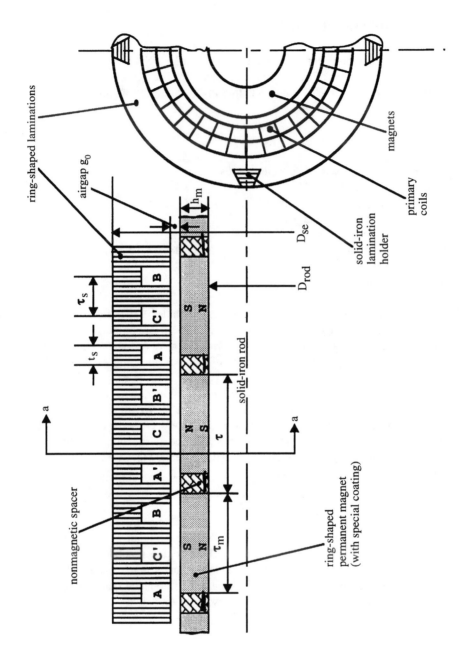

FIGURE 4.23
Tubular LPMSA geometry.

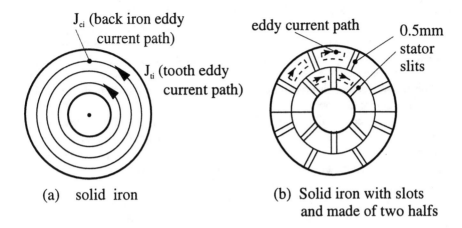

(a) solid iron

(b) Solid iron with slots
and made of two halfs

FIGURE 4.24
Solid iron stator with reduced core losses.

It follows from (4.50) and (4.51) that the current and the flux in the machine require approximately $t_1 f_1 = 2.882 \times 3.6 = 10.37$ periods over the travel along one stroke length. Because the current buildup could take place in milliseconds, we may assume steady-state operation for design purposes.

The low-frequency operation is a great benefit, as the core losses will be low although the PM flux density is rather high, and the core flux density is considerably high to reduce the core volume. At such a low frequency (of 3.6 Hz) the depth of penetration of the flux in the iron is given by

$$\delta_{iron} = \sqrt{(2/\mu_i \omega \sigma_i)}$$

$$= \sqrt{(2/200 \times 4\pi \times 10^{-7} \times 2\pi \times 3.6 \times 5 \times 10^6)} \tag{4.52}$$

$$= 8.39 \times 10^{-3} \text{m} = 8.39 \text{ mm}$$

As will be shown later the slot depth is about 4.5 mm, which compares with δ_{iron} obtained in (4.52) for a high degree of saturation ($\mu_i = 200\mu_0$). However, in order to use a solid iron stator core, we must reduce the "apparent" conductivity σ_i of the iron. To do this, we consider the eddy current paths in iron, which tend to be circular, as shown in Fig. 4.24. Making the stator core of two halves will facilitate the insertion of the coils into slots. This combined with thin (0.5-mm) radial slits in the teeth and back-iron core (Fig. 4.24) reduces the core losses by a factor of 8 to 10. Notice the eddy current paths in the slitted stator core made in two halves, shown in Fig. 4.24(b), as compared with that in a solid cylindrical core, shown in Fig. 4.24(a). On the other hand, for the laminated core, the coils may be inserted in the slots without splitting the stator core into two halves, as the entire core will have to be built tooth by tooth (say on a rod) after the coils are inserted in the slots. The laminated structure produces low core losses. But as the laminations are circular, at least in the back iron, leakage fluxes traverse the space between laminations. Three slots per pole and $\tau = 3$ cm imply an "additional" 1-mm airgap, which does not occur in a stator with solid iron. The ultimate choice between the two cores is determined by manufacturing costs. For design considerations there is no difference between the two cores. So in the following we consider a laminated core.

The PMs placed on the rotor solid iron rod (or shaft) are interleaved with nonmagnetic spacers and are coated with 0.1- to 0.2-mm material of great toughness. Otherwise, the magnets will wear out. A nonmagnetic stainless steel (0.1- to 0.2-mm) thin sleeve would be suitable. For bearings which are 0.5 m (or about 20 in.) apart, a mechanical airgap of 0.8 mm is required.

As we have chosen $\tau = 0.03$ m, from experience we choose a single-layer stator winding having one slot/pole per phase ($q = 1$). This choice results in a trapezoidal mmf distribution. Also, because the mover PMs produce (approximately) a trapezoidal airgap flux density, 120° rectangular current control is necessary to reduce the thrust pulsations. Assuming instantaneous commutation, only two phases will be conducting at a time. Surface PMs having $\tau_p = (5/6)\tau = 25$ mm (in length) have been found adequate to develop the desired thrust. It may be shown that the field due to the armature mmf is much lower than the PM field, and thus will not affect significantly the stator teeth saturation. To avoid excessive magnetic saturation and provide mechanical strength the minimum diameter of the mover (central) shaft should be D_{rod} = 18 mm. Samarium cobalt (SmCo$_5$) is used to avoid corrosion and for better temperature tolerance. For such magnets we have $B_r = 1.02$ T and $H_c = 0.732$ MA/m at 26 MGOe. With a high B_{g0} (close to B_r), the thickness of the magnets increases, teeth become thicker, and the slots thinner to reduce saturation. If the slot depth remains unchanged, the core back iron does not change for a given external diameter of the stator. But with a higher airgap flux density saturation of the stator and mover, core back iron increases appreciably. Consequently, there will be a degradation in the performance of the actuator.

In view of the proceeding discussion, we choose $B_g = 0.6$ T, with $B_r = 1.0$ T, $H_c = 0.7$ MA/m, and allow the temperature in the machine to be 125°C. We now calculate the PM ring height h_m (Fig. 4.25) from

$$H_m h_m + \frac{B_g}{\mu_0} k_c g (1 + k_s) = 0 \tag{4.53}$$

where k_s accounts for saturation and additional airgap between the stator laminations. We choose $k_s = 1$. Including the coating (or sleeve), total airgap $g = 0.8 + 0.2 = 1.0$ mm. We determine the Carter coefficient k_c in (4.53) from

$$k_c = \frac{1}{1 - \gamma g/\tau_s} \; ; \quad \tau_s = \text{slot pitch} = \frac{\tau}{3} \tag{4.54}$$

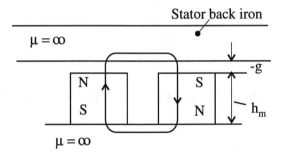

FIGURE 4.25
Line integral for calculating magnet height.

where

$$\gamma = \frac{(t_s/g)^2}{5 + (t_s/g)} \quad \text{for } t_s = \frac{\tau}{3} = 10 \text{ mm} \tag{4.55}$$

As the permeability of the PM is μ_0 at 125°C, the magnetic airgap of the motor is

$$g_m = h_m + g \tag{4.56}$$

Although this airgap should be considered in calculating k_c from (4.54), we will use g, as h_m is not known. Now, let slot width/slot pitch $= t_s/\tau_s/2 = 10/2 = 5$ mm. From (4.55)

$$\gamma = \frac{(5/1)^2}{5 + (5/1)} = 2.5 \tag{4.57}$$

Thus, (4.54) yields

$$k_c = \frac{1}{1 - (2.5 \times 1/10)} = 1.33 \tag{4.58}$$

To solve (4.53) we need the demagnetization characteristic of the PM, as given by

$$B_m = B_r + \mu_m H_m ; \quad H_m < 0 \text{ and } \mu_m = 1.13 \, \mu_0 \tag{4.59}$$

From (4.53) and (4.59) we obtain

$$H_m h_m = - \frac{0.6 \times 1 \times 10^{-3} \times 1.33(1 + 1)}{4\pi \times 10^{-7}} = -1.27 \times 10^3 \text{ A} \tag{4.60}$$

and,

$$B_m = B_{g0} = 0.6 = 1.0 + \mu_m H_m \qquad (4.61)$$

Thus,

$$H_m = \frac{-1.0 + 0.6}{1.13 \times 4\pi \times 10^{-7}} = -0.2817 \times 10^6 \text{ A/m} \qquad (4.62)$$

Combining (4.60) and (4.62) yields

$$0.2817 \times 10^6 \, h_m = 1.27 \times 10^3$$

$$\text{or} \quad h_m = 4.5 \text{ mm} \qquad (4.63)$$

We may now calculate the magnetic airgap g_m as seen by the armature mmf and by the PM. This g_m is given by

$$g_m = \mu_0 \frac{h_m}{\mu_m} + g k_c (1 + k_s)$$

$$= \frac{4.53}{1.13} + 1 \times 1.33 \, (1 + 1) = 6.74 \text{ mm} \qquad (4.64)$$

This value is rather large and indicates why armature reaction field is expected to be low.

The stator bore D_{si} is

$$D_{si} = D_{rod} + 2h_m + 2g = 18 + 2 \times 4.5 + 2 \times 1 = 29 \text{ mm} \qquad (4.65)$$

The external diameter of the stator, D_{se}, is constrained to 47 mm. Therefore, the stator slot height h_s added to the stator back-iron thickness h_{se} yields

$$h_s + h_{sc} = (D_{se} - D_{si})/2 = (47 - 29)/2 = 9 \text{ mm} \qquad (4.66)$$

Assuming the stator back-iron core flux density $Bsc = 1.37$ T, the back-iron thickness h_{sc} is given by

$$h_{sc} = \frac{B_{g0}}{B_{sc}} \left(\frac{\tau_p}{2} \right) = \frac{0.6}{1.3} \times \frac{30 \times 5/6}{2} = 5.7 \text{ mm}$$

However, as D_{se} is considerably greater than D_{si}, the PM flux spreads. So, the back-iron thickness may be reduced accordingly, and

$$h'_{sc} = h_{sc}[D_{si}/(D_{se} - h_{sc})]$$

$$= 5.77 \ [29/(47 - 5.77)] = 4.06 \text{ mm} \tag{4.67}$$

Therefore, the useful slot depth is given by

$$h_{su} = (h_s + h_{sc}) - h'_{sc} - 0.94$$

$$= 9 - 4.06 - 0.94 = 4.0 \text{ mm} \tag{4.68}$$

The distance 0.94 mm (chosen from experience) will remain unfilled, and thus mechanical contact between the coils on the stator and the moving magnets will be avoided. The unfilled space also permits a thin protective coating of the stator bore to avoid corrosion. This coating will also reduce the mechanical airgap from 0.8 mm to about 0.7 mm. [*Note*: The coating must be able to withstand the 125°C temperature and must be a good electrical (and magnetic) insulator.]

We now check the flux density B_{cr} in the rotor rod from

$$\frac{t_p}{2} \pi (D_{rod} + h_m) \ B_g = B_{cr} \frac{\pi}{4} D_{rod}^2 \tag{4.69}$$

or,

$$B_{cr} = \frac{30 \times 5}{2 \times 6} (18 + 4.5) \ 0.6 \times \frac{4}{18^2} = 2.06 \text{ T} \tag{4.70}$$

For this high-level flux density, the magnetic field $H = 38,000$ A/m. Consequently, the relative permeability is

$$\mu_r = \frac{2.06}{38,000 \times 4\pi \times 10^{-7}} = 43 \tag{4.71}$$

This high-level flux density can be reduced by using iron shunting rings, as shown in Fig. 4.26. The thickness of the ring is only 2 mm to avoid an appreciable armature reaction. If the shunting ring is not used and the leakage flux between the PMs is neglected, the 2.06-T flux density will distribute between the corners of the neighboring magnets along a 5-mm length. With a relative permeability of 43, this 5 mm of iron translates into 5/43 = 0.1163 mm of airgap, which is less than 1.75% of the total magnetic airgap g_m. Finally, as the Carter coefficient is 1.33, which allows for

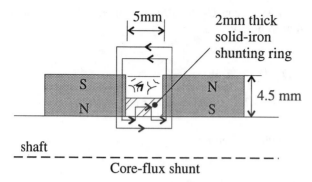

FIGURE 4.26
Core-flux shunt in PM mover.

the effective airgap, in the interest of cost the iron rings between the magnets may be omitted.

Thrust and Stator Slot mmf Calculations

For thrust calculations, we consider that only two phases conduct at a time and the current waveform is rectangular (120° wide), as shown in Fig. 4.27. The developed thrust is calculated in relation to Fig. 4.27, which shows that at a certain instant two stator-slot mmf per pole are under the influence of the PM field. Thus, the thrust per pole, F_{xp}, is

$$F_{xp} = B_{g0} 2n_c I(\pi D_{si}) = 0.6 \times 2 \times \pi \times 0.29 n_c I = 0.1093 n_c I \tag{4.72}$$

where n_c is the number of conductors and I the current. If p is the number of pole pairs, the thrust F_x for the entire machine is

$$F_x = 2pF_{xp} = 0.2186 n_c Ip \tag{4.73}$$

Before proceeding, we calculate the mover weight m, which is given by

$$m = \frac{\pi}{4}(D_{rod} + 2h_m)^2 \, 2p\tau\rho_{av} = \frac{\pi}{4} \times 0.027^2 \times 2p \times 0.03 \times 7 \times 10^3$$

$$= 0.24p \ kg \tag{4.74}$$

Assume that the pump plunger has the same weight, m. So total weight of mover is $2m = 0.48p$ kg. For a reverse acceleration $a = 0.2$ m/s² at the rated load, the dynamic thrust F_d becomes

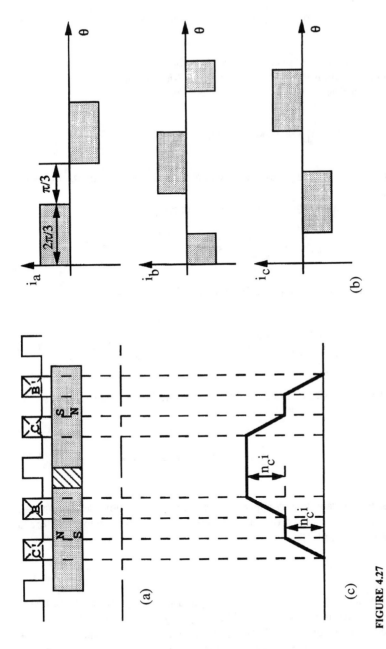

FIGURE 4.27
Armature mmf: (a) phases *b* and *c* conducting; (b) ideal current waveforms; and (c) mmf's of phases *b* and *c*.

$$F_d = 2ma = (0.48p)0.2 = 0.096p \quad \text{N} \tag{4.75}$$

Now, the actuator has to develop a maximum thrust $F_L = 227$ N. The design is exercised for this maximum thrust. We choose a 0.48-m 16-pole actuator, for which equating the forces yields

$$0.2186 n_c I_m p = 0.096p + 9.81 \times 0.48p + 227 \tag{4.76}$$

where the additional $9.81 \times 0.48\ p$ term, corresponding to about 20% thrust, allows for various adjustments.

With $p = 8$ (pole pairs), (4.76) yields (approximately)

$$(n_c I_m)_{slot} = 150 \text{ At} \tag{4.77}$$

Correspondingly, the maximum current density in a slot becomes

$$(J_{c0})_{max} = \frac{n_c I_m}{h_{su} t_s k_{fill}} = \frac{150}{4 \times 10^{-3} \times 5 \times 10^{-3} \times 0.45} = 16.6 \times 10^6 \text{ A/m}^2 \tag{4.78}$$

where h_{su} is slot depth, t_s slot opening, and k_{fill} fill factor. The rms current density is

$$(J_{c0})_{rm} = \sqrt{\frac{2}{3}} \ (J_{c0})_{max} = \sqrt{\frac{2}{3}} \times 16.6 \times 10^6 = 13.5 \times 10^6 \text{ A/m}^2 \tag{4.79}$$

Notice that this current density appears to be high; we will show that for a small mmf, the heat could be dissipated over the external surface of the stator and forced cooling will not be required.

Losses

Many of the losses in the motor are core losses p_{core} and winding losses p_w. Core losses are commonly expressed in terms of sinusoidally varying flux densities. However, in the present case the flux density in the stator teeth rises rapidly as the edge of the magnet crosses the tooth and then remains almost constant over the tooth width. So we consider the flux density variation as dB/dt, and express the core losses (approximately) as

$$p_c = 2k_l \left(\frac{dB}{dt}\right)^2 \text{ W/m}^3 \tag{4.80}$$

where k_l is a loss coefficient. As a first approximation, the time Δt required for the tooth flux density B_t to rise from zero to $B_{tooth} = 2B_{g0} = 1.2$ T is considered equal to the time during which the PM edge travels the tooth width. Now,

$$t_{tooth} = \tau_s - t_s \tag{4.81}$$

such that

$$\Delta t = \frac{\tau_s - t_s}{u_s} = \frac{\tau_s - t_s}{2\tau f_1} \tag{4.82}$$

and

$$\frac{dB_1}{dt} = \frac{B_{tooth}}{\Delta t} = \frac{B_{tooth}(2\tau f_1)}{\tau_s - t_s} \tag{4.83}$$

This change occurs four times per period, as shown in Fig. 4.28, for a fraction of time $4\Delta t/T$:

$$\frac{4\Delta t}{T} = \frac{4(\tau_s - t_s)}{2\tau f_1(1/f_1)} = \frac{2}{\tau}(\tau_s - t_s) \tag{4.84}$$

For the rest of the time the tooth core losses are zero. Thus, the tooth core losses per unit volume are

$$P_{ct} = 2k_l \frac{4\Delta t}{T} \left(\frac{dB_1}{dt}\right)^2 = 16k_l \frac{\tau f_1^2 B_{tooth}^2}{\tau_s - t_s} \tag{4.85}$$

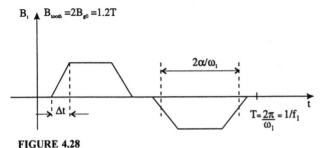

FIGURE 4.28
Stator tooth flux density vs. time.

For $q = 1$, slot width $= (\tau_s - t_s) =$ tooth width $= \tau_s/2 = \tau/6$, (4.85) becomes

$$P_{ct} = 96 k_I f_1^2 B_{tooth}^2 \quad \text{W/m}^3 \tag{4.86}$$

The volume of the stator tooth is

$$V_{tooth} = \frac{\pi}{4} \left[(D_{si} + 2h_s)^2 - D_{si}^2 \right] (\tau_s - t_s) \tag{4.87}$$

And the total core losses in a tooth become

$$(P_{ct})_{total} = P_{ct} V_{tooth} \quad \text{W} \tag{4.88}$$

The back-iron core flux density changes sign from $+B_{ccs}$ to $-B_{ccs}$ over the magnet length (Fig. 4.28) such that

$$\frac{dB_c}{dt} = \frac{2B_{ccs}}{(\tau_p/\tau)(f_1/2)} = \frac{2u_s B_{ccs}}{\tau_p} \tag{4.89}$$

which holds for the fraction of time

$$2f_1 \left(\frac{\tau_p}{\tau} \right) \frac{1}{2f_1} = \frac{\tau_p}{\tau} \tag{4.90}$$

Therefore, the stator back-iron (yoke) losses per unit volume are

$$P_{ccs} = 2k_I \left(\frac{dB_{ccs}}{dt} \right)^2 \frac{\tau_p}{\tau} = 32k_I \frac{\tau}{\tau_p} f_1^2 B_{ccs}^2 \tag{4.91}$$

The core volume between two teeth is

$$V_{ccs} = \frac{\pi}{4} \left[D_{se}^2 - (D_{si} + 2h_s)^2 \right] \tau_s \quad \text{m}^3 \tag{4.92}$$

And the losses in the yoke between two teeth become

$$(P_{ccs})_{xxxx} = P_{ccs} V_{ccs} \qquad (4.93)$$

Notice that the tooth flux density variation is faster but lasts for a shorter time, whereas the core flux density variation is slower and lasts longer.

Parameters as Functions of Number of Conductors per Slot (n_c)

The motor is considered a nonsalient pole synchronous machine with a large magnetic airgap g_m. The parameters R_s, the phase resistance, and L_s, the synchronous inductance per phase, are given in the following. We also determine the motional voltage E_l.

The per phase resistance R_s is given by

$$R_s = \frac{\rho\pi(D_{si} + h_s)n_s wp}{I_m/J_{comax}} = \frac{\rho\pi(D_{si} + h_s)J_{comax} \; n_s \; 2p \, n_c^2}{I_m n_c} \qquad (4.94)$$

Substituting numerical values in (4.94) yields

$$R_s = \frac{2.3 \times 10^{-8}\pi(0.029 + 0.005) \; 16.6 \times 10^6 \times 16 n_c^2}{150} = 0.435 \times 10^{-2} n_c^2 \qquad (4.95)$$

The synchronous inductance L_s consists of the slot leakage inductance $L_{s\sigma}$ and the magnetizing inductance L_m. These are, respectively, given by

$$L_{s\sigma} = \mu_0 n_c^2 \; 2p\pi(D_{si} + h_s) \left[\frac{h_{su}}{3t_s} + \frac{h_{si}}{t_s} \right] \qquad (4.96)$$

and

$$L_m = \frac{6\mu_0}{\pi^2}(n_c p)^2 \; \frac{\tau\pi D_{si}}{p g_m} \qquad (4.97)$$

Substituting numerical values in (4.96) and (4.97) we obtain

$$L_{s\sigma} = 4\pi \times 10^{-7} \times 16\pi(0.029 + 0.005) \left[\frac{4}{15} + \frac{1}{5} \right] n_c^2 = 1 \times 10^{-6} n_c^2 \qquad (4.98)$$

and

$$L_m = \frac{6 \times 4\pi \times 10^{-7}}{\pi^2} \times \frac{8^2 \times 0.03 \times \pi \times 0.029}{8 \times 6.74 \times 10^{-3}} n_c^2 = 2.48 \times 10^{-6} n_c^2 \qquad (4.99)$$

Adding (4.93) and (4.99) yields

$$L_s = L_{s\sigma} + L_m = (1 + 2.48)10^{-6} n_c^2 = 3.48 \times 10^{-6} n_c^2 \qquad (4.100)$$

Finally, the stator time constant is

$$\tau_e = \frac{L_s}{R_s} = \frac{3.48 \times 10^{-6} n_c^2}{0.435 \times 10^{-2} n_c^2} = 0.8 \text{ ms} \qquad (4.101)$$

The motional line voltage E_l has a trapezoidal waveform, which can be approximated by a rectangular waveform. Thus, at the rated (synchronous) speed u_s, E_l is obtained from

$$E_l = B_g u_s (\pi D_{si}) \, 2p \times 2n_c \qquad (4.102)$$

or,

$$E_l = 0.6 \times 0.216 \times \pi \times 0.029 \times 16 \times 2n_c = 0.378 n_c \qquad (4.103)$$

In order to determine n_c and the wire size, we study the controller in some detail.

Rectangular Current Control

A typical rectangular current control system is shown in Fig. 4.29. The control system includes a voltage-source PWM transistor inverter and a protection/braking resistor. The bipositional switch is turned on by the threshold voltage V^*. The braking resistor is common to PWM converters in the kilowatt power range and is a part of the converter. A six-element-per-pole-pair pitch position transducer, fixed at one end of the stator, is used for the commutation of the phases in the PWM inverter. The elements of the position transducer are shifted by $\tau/3 = 10$ mm. Each element turns on and turns off a single power transistor to provide a $120°$ conducting period. The PWM for current control during this conduction interval is done by the PWM circuit via an independent current hysteresis or ramp controller for all phases.

It follows from Fig. 4.30 that ideally only two transistors conduct at a time. For example, the position sensors $P - T_1$ and $P - T_6$ produce, respectively, positive and negative voltages (currents) in phase a; $P - T_3$ and $P - T_4$ in phase b; and $P - T_5$ and $P - T_2$ in phase c. Thus, the stator mmf jumps every $60°$. We now locate the position sensor which fires the transistor T_1 $90°$ (or $\tau/2$) behind the axis of phase a (with respect to the direction of motion). The power angle in the motor varies from $60°$ to

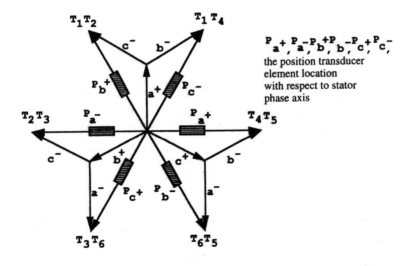

$P_a{}^+, P_a{}^-, P_b{}^+, P_b{}^-, P_c{}^+, P_c{}^-,$
the position transducer
element location
with respect to stator
phase axis

(a)

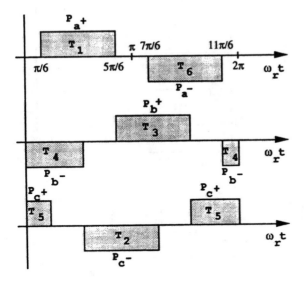

FIGURE 4.29
Rectangular current controller.

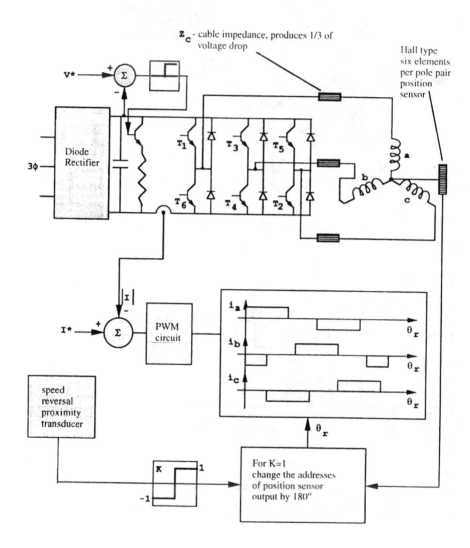

FIGURE 4.30
Six-element position sensor.

120°, with an average of 90°. To reverse the direction of motion we simply have to change the power angle by 180°. This is done by switching the addresses (transistors) turned on and off by a position sensor element by 180° such at a^+ becomes a^-, b^-, and c^+ becomes c^-. The command for speed reversal should come from a proximity transducer fixed to the stator on both sides to signal the point from which the regenerative braking and speed reversal should begin. This will result only in the above change of the addresses in the position sensor outputs: $a^+ \rightarrow a^-$; $b^+ \rightarrow b^-$; and $c^+ \rightarrow c^-$, and vice versa.

The frequency of the LPMSA sustained oscillations is controlled through the reference current I^* manually or by a separate controller, if necessary. To control the current level in the motor an on-off controller is used.

On-Off Current Controller

Consider the inverter motor equations for on-off time intervals (Fig. 4.31). During on-time, two transistors T_{A+} and T_{B-} conduct. Thus, the current flows through phases a^+b^- connected in series. The line-induced voltage is constant during this interval, and the voltage equation becomes

$$V_0 = 2R_s i + 2L_s \frac{di}{dt} + E_l \qquad (4.104)$$

The current is allowed to vary from I_{0min} to I_{0max}, to provide hysteresis control. Let the steady state be achieved at $t = 0$ with $i = I_{0min}$. The solution to (4.104) becomes

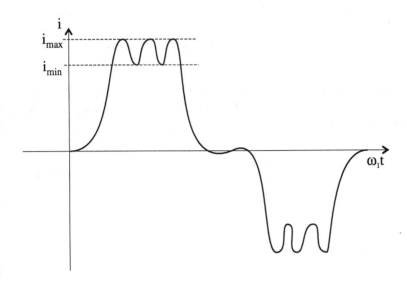

FIGURE 4.31
An on-off controller.

$$i(t) = \frac{V_0 - E_l}{2R_s} (1 - e^{-tR/L_s}) + I_{0min} e^{-tR/L_s}, \qquad 0 \le t \le t_{on} \qquad (4.105)$$

For the current rise, we must have $V_0 > E_l$; that is, the machine is underexcited.

During the off-time interval, both T_{A+} and T_{B-} are turned off. The diodes D_{A-} and D_{B+} provide for the path to charge the filter capacitor C_f. Open-switch S signifies that the power supply does not accept power retrieval. The circuit equation now becomes

$$V_0 = 2R_s + 2L_s \frac{di}{dt} - E_l + \frac{1}{C_f} \int i\, dt \qquad (4.106)$$

Let the turn-off process begin at $t_i = t_{on}$ (with $t' = t - t_{on}$) with $i = I_{0max}$. The solution to (4.106) is

$$i(t) = -\frac{(E_l + V_0)}{2\omega L_s} e^{-\alpha t'} \sin \omega t - I_{0max} \frac{\omega_0}{\omega} c^{-\alpha t'} \sin(\omega t - \phi) \qquad (4.107)$$

where

$$\omega_0 = (\sqrt{(L_s C_f/2)}\,); \qquad \alpha = R_s/2L_s; \qquad \omega = \sqrt{\omega_0^2 - \alpha^2}$$

and

$$\phi = \tan^{-1}(\omega/\alpha)$$

The PWM circuit will adjust t_{on} and t_{off} to comply with the hysteresis band ($I_{0max} - I_{0min}$).

To avoid current oscillations, we restrict the turn-off time to less than 90° at the frequency ω. We choose first a maximum switching frequency of 10 kHz, as the electrical time constant is 0.8 ms, from (4.101). So the smallest off-time is 0.1 ms, which corresponds to one-fourth of $2\pi/\omega$. Thus,

$$\frac{1}{4}\left(\frac{2\pi}{\omega}\right) = 0.1 \text{ ms} \qquad (4.108)$$

Solving (4.108) yields

$$\omega = 1.57 \times 10^4 \text{ r/s} \qquad (4.109)$$

Now, $\alpha = R_s/2L_s = 625$

So,

$$\omega_0 = \sqrt{\omega^2 + \alpha^2} = \sqrt{[(1.57 \times 10^4)^2 + 625^2]} = 1.57 \times 10^4 = \sqrt{(L_s C_f / 2)}$$

Consequently,

$$C_f = \frac{1}{2\omega_0^2 L_s} = \frac{1}{2(1.57 \times 10^4)^2 3.48 \times 10^{-6} n_c^2} = \frac{5.8 \times 10^{-4}}{n_c^2} \qquad (4.110)$$

We now use (4.106) with $i(t_{on}) = I_{0max}$ and $t = t_{on}$. For a three-phase diode rectifier, the ideal output voltage is

$$V_0' = \frac{3}{\pi} V_m = \frac{3}{\pi} \times 220 \sqrt{2} = 297 \text{ V} \qquad (4.111)$$

Assuming a 97-V drop in the feeding cable, the dc link voltage $V_0 = 297 - 97 = 200$ V.

Now, we have to fix I_{0max}, I_{0min}, and t_{on} to determine n_c. The fastest current variation occurs at standstill, when $E_l = 0$. Therefore, from (4.105) with $I_{0max} = 1.1 I_m$, $I_{min} = 0.9 I_m$, and $t_{on} = 0.1$ ms we obtain, upon multiplying by n_c,

$$(1.1 - 0.882 \times 0.9) n_c I_m = \frac{200 n_c}{2 \times 4.35 \times 10^{-3} n_c^2} (1 - 0.882) \qquad (4.112)$$

Solving (4.112) with $n_c I_m = 150$ we obtain

$$n_c = 59 \text{ conductors/slot} \qquad (4.113)$$

The motional voltage at maximum speed is from (4.103)

$$E_{tm} = 0.378 \times 59 = 22.3 \text{ V} \qquad (4.114)$$

We must now ascertain that at this maximum voltage the t_{on} time is only slightly higher but easily achievable, as $V_0 - E_{tm} = 220 - 22.3 = 177.7$ V. Let us try a switching frequency of 2.5 kHz and $t_{on} = 0.4$ ms. Proceeding as above yields

$$(1.1 - 0.9 e^{-0.4/0.8}) n_c I_m = \frac{200 n_c}{2 \times 4.35 \times 10^{-3} n_c^2} (1 - e^{-0.4/0.8}) \qquad (4.115)$$

from which

$$n_c = 133 \text{ conductors/slot} \tag{4.116}$$

and

$$E_{tm} = 0.378 \times 133 = 50 \text{ V} \tag{4.117}$$

Also,

$$I_m = \frac{150}{133} = 1.13 \text{ A} \tag{4.118}$$

For $n_c = 133$, the stator resistance is, from (4.95),

$$R_s = 0.435 \times 10^{-2}(133)^2 = 77 \ \Omega \tag{4.119}$$

and

$$V_0 - 2R_sI_m = 200 - 2 \times 77 \times 1.13 = 26 \text{ V} \tag{4.120}$$

However, to maintain a constant current at the maximum speed, $(V_0 - 2R_sI_m)$ must exceed the motional voltage of 50 V, given by (4.118). So we now choose a switching frequency of 5 kHz (which is common to IGBTs). Thus, we obtain

$$(1.1 - 0.9e^{-0.2/0.8})n_cI_m = \frac{200 \ n_c}{2 \times 4.35 \times 10^{-3} \ n_c^2} (1 - e^{-0.2/0.8}) \tag{4.121}$$

which yields

$$n_c = 85 \ \text{ conductors/slot} \tag{4.122}$$

and

$$E_{tm} = 0.378 \times 85 = 32.13 \text{ V} \tag{4.123}$$

$$R_s = 0.435 \times 10^{-2} (85)^2 = 31.42 \ \Omega \tag{4.124}$$

$$I_m = 150/85 = 1.76 \text{ A} \tag{4.125}$$

Finally,

$$V_0 - 2R_sI_m = 200 - 2 \times 31.42 \times 1.76 = 81.1 \text{ V} \tag{4.126}$$

The above values, (4.122) through (4.126) are considered adequate. To determine C_f we proceed as follows: At 5 kHz,

$$\omega = \frac{2\pi f}{r} = \frac{2\pi \times 5 \times 10^3}{4} = 7.85 \times 10^3 \text{ r/s} \tag{4.127}$$

$$\omega_0 = \sqrt{[(7.85)^2 + (1/2 \times 0.8 \times 10^{-3})^2]} = 7.875 \times 10^3 \text{ r/s} \tag{4.128}$$

and

$$C_f = 1/[2(7.875 \times 10^3)^2 \times 3.48 \times 10^{-6} \times 85^2] = 0.32 \ \mu\text{F} \tag{4.129}$$

In general, the filter has a much larger capacitance so that ω_0 will be smaller, leading to a smaller ω and thereby avoiding oscillations.

The major results of this section are as follows: number of conductors per slot $n_c = 85$ and maximum current $I_m = 1.76$ A. This will enbale us to choose the rating of the converter. Power transistors of 5 A and 1000 V rating may be used in the inverter. The motor is to be operated with a 220-V, three-phase voltage source. The maximum switching frequency of 5 kHz is adequate.

Wire Size

The area of cross section of the wire making the stator coils is given by

$$A_{c0} = \frac{I_m}{J_{comax}} = \frac{1.76}{16.6 \times 10^6} = 0.106 \text{ mm}^3 \tag{4.130}$$

which corresponds to a wire of 0.367 mm (or 0.0144 in.) diameter. So we choose AWG-27 wire which has a bare diameter of 0.0142 in. The fill factor for this wire is

$$k_{fill} = \frac{\text{conductor area}}{\text{slot area}} = \frac{\pi}{4} \frac{(0.361)^2}{5 \times 4 \times 10^{-3}} = 0.5 \tag{4.131}$$

which is greater than the originally assumed value but is still acceptable.

Since we have chosen 85 conductors per slot, each coil must have 85 turns. It is expected that the actuator will produce a 227-N thrust at 38% duty cycle. For this case the winding losses will be about 73 W and the wire temperature will not exceed 125°C.

Peak Thrust Density

From (4.76) the peak thrust is

$$F_{xk} = 0.096 \times 8 + 9.81 \times 0.48 \times 8 + 227 = 265.44 \text{ N} \tag{4.132}$$

Thus, the peak thrust density is given by

$$f_{xk} = \frac{F_{xk}}{\pi(D_{rod} + 2h_m)\,2p\tau} = \frac{265.44}{\pi \times 0.027 \times 0.48} = 0.652 \text{ N/cm}^2 \quad (4.133)$$

Considering that the mover external diameter is only 27 mm the value of the thrust density given by (4.133) is rather large.

REFERENCES

1. T. J. E. Miller, *Brushless permanent-magnet and reluctance motor drives* (Clarendon Press, Oxford, 1989).
2. S. A. Nasar, I. Boldea, and L. E. Unnewehr, *Permanent magnet, reluctance and self-synchronous motors* (CRC Press, Boca Raton, FL, 1993).
3. J. F. Eastham, "Novel synchronous machines: Linear and disc, *Proc. IEE*(B), vol. 137, 1990, pp. 49-58.
4. Z. Deng, I. Boldea, and S. A. Nasar, "Fields in permanent magnet linear synchronous machines," *IEEE Trans.*, vol. MAG-22, 1986, pp. 107-112.
5. Z. Deng, I. Boldea, and S. A. Nasar, "Forces and parameters of permanent magnet linear synchronous machines," *IEEE Trans.*, vol. MAG-23, 1987, pp. 305-309.
6. G. Xiong and S. A. Nasar, "Analysis of fields and forces in a linear synchronous machine, based on the concept of magnetic charge," *IEEE Trans.*, vol. MAG-25, 1989, pp. 2713-2719.
7. T. Nizuno and H. Yamada, "Performance characteristics of a transfer machine using permanent magnet type linear synchronous motor," *Rec. ICEM*, 1992, vol. 2, pp. 721-725.
8. K. Yoshida, N. Teshina, and E. Zen, "FEM analysis of thrust and lift forces in controlled permanent magnet LSM," *Rec. ICEM*, 1990, vol. 3, pp. 1100-1106.
9. I. Boldea, S. A. Nasar, and Z. Fu, "Fields, forces and performance of air-core linear self-synchronous motor with rectangular current control," *IEEE Trans.*, vol. MAG-24, 1988, pp. 2194-2203.
10. J. J. Cathey, D. A. Topmiller, and S. A. Nasar, "Performance of a tubular self-synchronous motor for artificial heart pump drive," *IEEE Trans.*, vol. BME-33, 1986, pp. 315-319.
11. K. Yoshida, Y. Tsubone, and T. Yamashita, "Propulsion control of controlled-PM LSM maglev carrier," *Rec. ICEM*, 1992, vol. 2, pp. 726-730.

CHAPTER 5

LINEAR RELUCTANCE SYNCHRONOUS ACTUATORS

In Chapter 3 we observed that the thrust/kVA and thrust/loss ratios for linear induction actuators (LIAs) are rather low. On the other hand, in Chapter 4 we concluded that these ratios are high for linear permanent magnet synchronous actuators (LPMSAs). These are relatively expensive, however, owing to the costs of magnets. Recent research in rotary reluctance machines has led to the development of linear reluctance synchronous actuators (LRSAs), which are relatively inexpensive and have good performance indices. Most LRSAs correspond to their rotary counterparts, including reluctance motors with traveling fields, switched reluctance motors, stepper motors, etc. LRSAs are especially suited for low-speed (less than 2 m/s) applications.

In analyzing an LRSA, it is to be noted that the winding concepts and details of LIAs (Chapter 3) are valid for LRSAs, and its speed being low, end effects in an LRSA may be neglected. However, the presence of high saliency on its secondary must be taken into account. In essence, theory pertaining to rotary reluctance motors is also applicable to LRSAs, except for the existence of normal forces in the latter, which merits special treatment.

5.1 PRACTICAL CONFIGURATIONS AND SALIENCY COEFFICIENTS

Like most linear electric actuators, LRSAs may be tubular or flat, single-sided or double-sided, with the secondary as the mover. The primary windings are three-phase ac windings as in LIAs. The secondary should have a high saliency. The simplest conventional salient-pole mover concept is shown in Fig. 5.1. A primary sinusoidal mmf aligned with the d-axis produces an airgap flux density distribution whose fundamental is B_{gd1}, as shown in Fig. 5.1(a). Similarly, when the mmf is aligned with the q-axis we obtain the distribution shown in Fig. 5.1(b), with B_{gq1} as the value of the fundamental component. The ratios between these values of the fundamentals and the airgap flux density fundamental B_{g1} produced by the same mmf in a uniform airgap configurations are termed saliency coefficients k_{dm1} and k_{qm1} such that

$$k_{dm1} = \frac{B_{gd1}}{B_{g1}} \; ; \quad k_{qm1} = \frac{B_{gq1}}{B_{g1}} \tag{5.1}$$

A very good assessment of the flux density distribution in the airgap in the presence of saturation can be obtained by the finite-element method (FEM). However, for a preliminary design, approximate analytical expressions may be used for flux den-

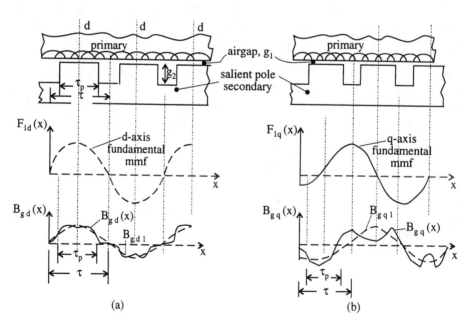

(a)   (b)

FIGURE 5.1

Airgap fields along (a) d-axis and (b) q-axis of a conventional synchronous machine.

sity determination. Neglecting fringing and saturation, approximate magnetic circuit analysis applied to Fig. 5.1 yields [1]

$$k_{dm1} = a + (1 - a)\left[\frac{\tau_p}{\tau} + \frac{1}{\pi} \sin\left(\frac{\pi}{\tau} \tau_p\right)\right] \tag{5.2}$$

$$k_{qm1} = a + (1 - a)\left[\frac{\tau_p}{\tau} - \frac{1}{\pi} \sin\left(\frac{\pi}{\tau} \tau_p\right)\right] \tag{5.3}$$

where $a = g_1/g_2$ and other symbols are shown in Fig. 5.1. It can be readily verified that for $g_2 = 20g_1$ (or $a = 0.05$) and $\tau_p/\tau = 0.5$, $k_{dm1} = 0.83$, $k_{qm1} = 0.22$, and the ratio $k_{dm1}/k_{qm1} = 3.73$. Compared with that of a round-rotor, or uniform airgap machine, the d-axis inductance of the salient-pole machine having $\tau_p/\tau = 0.5$ is about 17% smaller. If saturation is included in the analysis, the d-axis inductance will be reduced. Furthermore, fringing will tend to decrease the d-axis inductance, ultimately making the ratio k_{dm1}/k_{qm1} less than 3 for the above example. Thus, the configuration shown in Fig. 5.1 is not an ideal choice for the secondary of an LRSA.

In order to increase the salience ratio in rotary reluctance machines, segmented rotors [2] and axially laminated rotors have been proposed. Linear reluctance motors with segmented secondaries also have a relatively higher saliency ratio [3]. The segmented secondary essentially features one flux barrier per pole (Fig. 5.2). The segment does not have a zero magnetic potential. Its value is P for the q-axis field and zero for the d-axis field since the latter does not cross the barrier between the segments. The magnetic scalar potential P is obtained from the condition that the total flux entering AB [Fig. 5.2(b)] leaves through the secondary segment along BC. Consequently,

$$\frac{\mu_0 \tau}{\pi g_1} \int_{(1 - \tau_p/\tau)\pi/2}^{\pi/2} (F_{1q} \sin \theta - P)\, d\theta = \mu_0 \frac{Pb}{g_2}\ ; \qquad g_2 = \frac{1}{2}(\tau - \tau_p) \tag{5.4}$$

Solving for P yields

$$P = \left(\frac{\tau}{\pi g_1} F_{1q} \sin \frac{\pi \tau_p}{2\tau}\right) \Bigg/ \left(\frac{b}{g_2} + \frac{\tau_p}{2\pi g_1}\right) \tag{5.5}$$

Finally, based on the procedure given in Reference [1] we obtain

$$k_{dm1} = \frac{\tau_p}{\tau} - \frac{1}{\pi} \sin \pi \frac{\tau_p}{\tau} \tag{5.6}$$

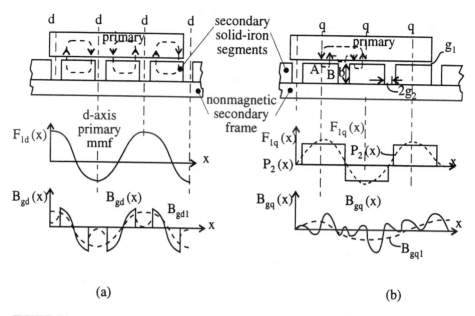

(a)   (b)

FIGURE 5.2
Airgap fields along (a) d-axis and (b) q-axis of a solid segmented pole synchronous machine.

$$k_{qm1} = \frac{\tau_p}{\tau} + \frac{1}{\pi} \sin \pi \frac{\tau_p}{\tau} - \frac{4P}{\pi F_{1q}} \sin \frac{\pi \tau_p}{2\tau} \qquad (5.7)$$

For the numerical values $\tau = 60$ mm, $\tau_p/\tau = 0.9$, $g_1 = 0.4$ mm, and $g_2 = 0.5\tau(1 - \tau_p/\tau)$ $= 60/2 \ (1 - 0.9) = 3$ mm, (5.5) through (5.7) give $P/F_{1q} = 0.636$, $k_{dm1} = 0.8016$, and $k_{qm1} = 0.1982$. In this case $k_{dm1}/k_{qm1} = 4.04$, which is slightly better than that for the conventional structure of Fig. 5.1.

From the above discussion we draw the following conclusions. First, topological changes cannot increase the saliency ratio above 5. Secondly, harmonic fields induce eddy current losses in the mover segments. Thirdly, for $k_{dm1} \approx 0.8$ and $\tau_p/\tau = 0.9$ the difference

$$L_{dm} - L_{qm} = (k_{dm1} - k_{qm1})L_{mj} \qquad (5.8)$$

is no greater than 0.6 L_m, L_m being the magnetizing inductance.

Compared with an induction actuator having an identical primary and the airgap, a linear reluctance actuator has a much lower thrust and a 25% increase in the magnetizing current, unless the saliency ratio is considerably increased. This goal can be accomplished in two ways. One approach is to stamp flux barriers in a conventional secondary lamination, as shown in Fig. 5.3(a). Saturable bridges retain the continuity of the laminations. At low loads (or small F_{1q}) the bridges are unsaturated and $k_{dm1} \approx$

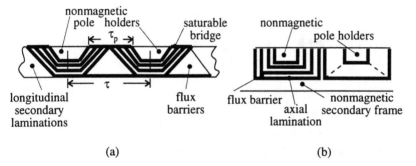

(a) (b)

FIGURE 5.3
Multiple flux barrier anisotropic secondary: (a) with stamped conventional laminations; (b) with transverse lamination-insulation layers.

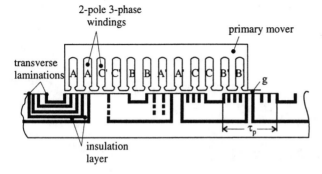

FIGURE 5.4
Flat single-sided LRSA with primary mover.

k_{qm1}. On load, F_{1q} increases, the bridges saturate and thereby k_{qm1} decreases. The ratio k_{dm1}/k_{qm1} depends on the bridge length/width ratio and on the number of flux barriers per pole. For practical reasons, it is not feasible to have more than 8 to 10 flux barriers in the secondary of an LRSM with a pole pitch of about 100 mm. Also, it is not advantageous to have more than 10 to 12 flux barriers [4] to increase the k_{dm1}/k_{qm1} ratio, although harmonic fields and thrust pulsations decrease with an increase in the number of flux barriers.

The alternative approach to obtaining a large saliency ratio in an LRSM is to use an axially laminated secondary, illustrated in Fig. 5.3(b). In this case thin lamination-insulation layers are used to fabricate the secondary. Ratios of k_{dm1}/k_{qm1} in excess of 25 have been thus obtained for rotary reluctance machines having pole-pitch/airgap ratios greater than 250 [5]. Therefore, a transversely laminated anisotropic (TLA) secondary, corresponding to its rotary counterpart, is expected to yield a high-performance LRSA. Three possible configurations of LRSAs with TLA secondaries are shown in Figs. 5.4 and 5.5, which have high saliency ratios. Even a higher saliencyis obtainable if (inexpensive bonded ferrite) magnet layers are used, as shown in Fig. 5.7 [6]. The following design guidelines can be formulated from a study of reluctance machines with axially laminated anisotropic (ALA) rotors: (i) keep the

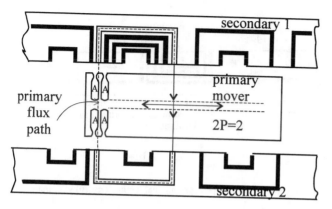

FIGURE 5.5
Flat double-sided LRSA with primary mover.

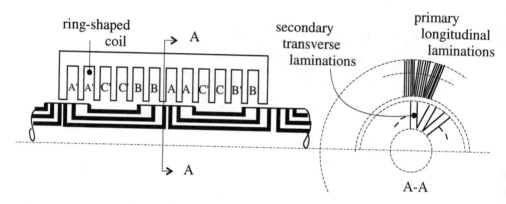

FIGURE 5.6
Tubular LRSA with secondary mover.

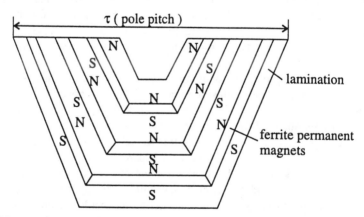

FIGURE 5.7
Secondary with transverse lamination-PM layers.

airgap as small as possible; (ii) keep the Carter coefficient below 1.4; and (iii) keep the number of slots/pole per phase slightly greater than 1. These considerations lead to double-layer windings for flat LRSAs and single-layer windings for tubular reluctance actuators.

5.2 FIELD DISTRIBUTIONS AND SALIENCY RATIO

A realistic study of the field distribution in a secondary structure with transverse lamination-insulation layers can be obtained via FEM [6]. From a study of field distributions in an ALA-rotor reluctance machine [7] it was found that a saliency ratio $k_{dm1}/k_{qm1} \approx 25$ could be obtained in a 24-slot 2-pole machine with slots/pole per phase of 4 and a pole-pitch/airgap ratio of 300. Also, as indicated earlier the ratio of lamination thickness to insulation thickness d_m/d affects the saliency ratio; and a value of d_m/d_i between 1.0 and 1.5 is considered practical [4,5]. This discussion is based on results obtained by FEM. Next, we consider an approximate analysis of the field distributions.

Analytical Approach to Field Distributions

In this approximate analysis, we consider two different airgap functions $g_d(x)$ and $g_q(x)$ to account for the lamination-insulation layers of the secondary as acted upon by the fundamental component of the primary mmf. These airgap functions are shown in Fig. 5.8, for which we have

$$g_d(x) = k_c g, \qquad 0 \le x \le \tau_p/2$$

$$= k_c g + \frac{\pi}{2}\left(x - \frac{\tau_p}{2}\right), \quad \tau_p/2 \le x \le \tau/2 \tag{5.9}$$

$$g_q(x) = k_c g + x d_i/(d_m + d_i), \qquad 0 \le x \le \tau_p/2$$

$$= \frac{\tau_p d_i}{2(d_m + d_i)} + \frac{\pi}{2}\left(x - \frac{\tau_p}{2}\right), \quad \tau_p/2 \le x \le \tau/2 \tag{5.10}$$

where k_c is Carter's coefficient and other symbols are defined in Fig. 5.8. We now assume that the q-axis flux density does not change sign under each pole. This assumption will lead to a higher q-axis flux density than in reality, resulting in a higher k_{qm1}. Also, we assume that saturation occurs only in the d-axis. Thus, the fundamentals of the airgap flux densities along the two axes are obtained from

$$B_{gd1} = \frac{4}{\tau}\mu_0 F_{1d} \int_0^{\tau/2} \frac{\cos^2(\pi x/\tau)}{g_d(x)} \, dx \tag{5.11}$$

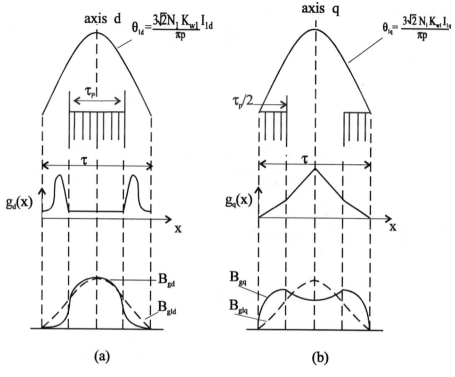

FIGURE 5.8
Airgap functions and flux densities along (a) d-axis and (b) q-axis.

and

$$B_{gq1} = \frac{4}{\tau} \mu_0 F_{1q} \int_0^{\tau/2} \frac{\sin^2(\pi x/\tau)}{g_q(x)} \, dx \qquad (5.12)$$

For a uniform airgap machine with no saturation we have

$$B_{g1} = \frac{\mu_0 F_{1d}}{k_c g} \qquad (5.13)$$

Consequently, (5.1) and (5.11) through (5.13) yield

$$k_{dm1} = \frac{4}{\tau} k_c g \int_0^{\tau/2} \frac{\cos^2(\pi x/\tau)}{g_d(x)} \, dx \qquad (5.14)$$

and

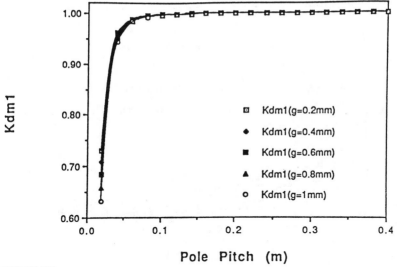

FIGURE 5.9
k_{dm1} versus pole pitch.

$$k_{qm1} = \frac{4}{\tau} k_c g \int_0^{\tau/2} \frac{\sin^2(\pi x/\tau)}{g_q(x)} dx \qquad (5.15)$$

These integrals are evaluated numerically, and the respective results of (5.14) and (5.15) are shown in Figs. 5.9 and 5.10.

We now apply the preceding results to LRSAs. First, in most cases, except for applications requiring very large thrusts, the pole pitch of an LRSA is likely to be less than 0.1 m and the airgap is expected to be about 0.6 mm. If we use permanent magnet (PM) layers (Fig. 5.7) in place of insulation layers, the airgap can be increased if so desired. Notice that the PMs tend to reduce the airgap flux density in the q-axis. Furthermore, the effective mmf's of PMs must be greater than the maximum q-axis primary mmf to avoid PM demagnetization. Finally, in line with the discussions up to this point, we arrive at the following conclusions regarding LRSAs:

(i) If the τ/g ratio is greater than 150, the τ_p/τ ratio can be kept below 2/3, for $k_{dm1} > 0.96$ and increases only slightly with an increase in τ_p/τ, while k_{qm1} increases considerably. The goal is to maximize k_{dm1}/k_{qm1} and (k_{dm1} / k_{qm1}).

(ii) To obtain a large $k_{dm1}/k_{qm1} \approx 25$, τ/g must be greater than 250, which is feasible only in high-thrust applications, and results in the best performance.

(iii) The performance (efficiency and power factor) deteriorates with a reduction in size and thrust of the LRSA.

(iv) Table 5.1 summarizes recommended values of pole-pitch airgap pairs for LRSAs. In this table, we consider $\tau = 15$ as the lower limit, as with slots/pole per phase, $q = 1$, the slot opening is 3 mm and tooth width 2 mm, which reflect realistic practical values.

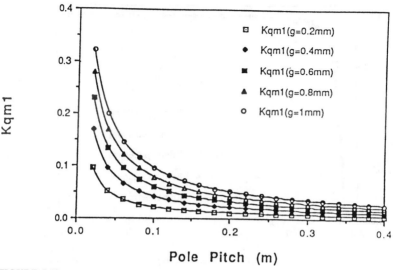

FIGURE 5.10
k_{qm1} versus pole pitch.

TABLE 5.1 Pole-pitch airgap pairs for LRSAs

Pole pitch τ, mm	100 to 60	60 to 30	30 to 15
Airgap g, mm	0.4 to 0.3	0.3 to 0.2	0.2 to 0.1

(v) Because of saturation in the d-axis, the unsaturated value of k_{dm1} decreases to k_{dm1}^s given by

$$k_{dm1}^s = \frac{k_{dm1}}{1 + k_{sd}} \tag{5.16}$$

where k_{sd} is the d-axis saturation factor, which should not be allowed to exceed a value of 0.4.

(vi) The ideal value of the ratio d_m/d_i is 1. However, in this case, the airgap flux density becomes lower. Practical values of d_m/d_i are between 1.0 and 1.5 [4,5,7].

(vii) The number of slots/pole per phase should be greater than 1.

5.3 MATHEMATICAL MODEL

With the longitudinal end effect neglected, the theory of rotary reluctance synchronous motors is applicable to LRSAs. Thus, the dq-space-phasor model in stator coordinates is described by the following set of equations [8]:

$$v_1 = R_s i_s + p\lambda_q - j\omega_r\lambda_s \tag{5.17}$$

$$v_1 = v_d + jv_q; \quad i_1 = i_d + ji_q; \quad \lambda_1 = \lambda_d + j\lambda_q \tag{5.18}$$

where $p = d/dt$, $\omega_r = u\pi/\tau$, and u is the speed of the mover. If the primary is the mover (as in Fig. 5.4) then u is in the direction opposite to that of the traveling field. The flux linkages λ_d and λ_q are defined by

$$\lambda_d = L_d i_d \quad \text{and} \quad \lambda_q = L_q i_q \tag{5.19}$$

The thrust F_x is obtained from

$$F_x u = \frac{3}{2}\omega_r \, \text{Re}(j\lambda_1 i_1^*) \tag{5.20}$$

which yields

$$F_x = \frac{3\pi}{2\tau}(L_d - L_q)i_d i_q \tag{5.21}$$

With PMs in the q-axis of the secondary (Fig. 5.7) the q-axis flux linkage modifies to

$$\lambda_q = L_q i_q - \lambda_{PM} \tag{5.22}$$

and the expression for the thrust becomes

$$F_x = \frac{3\pi}{2\tau}\left[\lambda_{PM} i_d + (L_d - L_q)i_d i_q\right] \tag{5.23}$$

Because λ_{PM} is kept greater than $L_q i_q$ as mentioned earlier, for F_x we have

$$F_x > \frac{3\pi}{2\tau}L_d i_d i_q \quad \text{for} \quad \lambda_{PM} > L_q i_q \tag{5.24}$$

Returning to the space-phasor equations, Park transformation relates dq-quantities to phase quantities such that

$$v_1 = \frac{2}{3}\left(v_a + v_b e^{j2\pi/3} + v_c e^{-j2\pi/3}\right)e^{j\theta_{er}} \tag{5.25}$$

with

$$\dot{\theta}_{er} = -\omega_r = -\frac{\pi u}{\tau} \qquad (5.26)$$

The normal (attraction) force is obtained from

$$F_{na} = \frac{3}{2} \frac{\partial}{\partial g} \left(\lambda_d i_d + \lambda_q i_q\right)_{i_d, i_q = const} \qquad (5.27)$$

Because λ_q and λ_{PM} do not vary with the airgap g, (5.27) may be approximated as

$$F_{na} = \frac{3}{2} i_d^2 \frac{\partial L_{dm}}{\partial g} \qquad (5.28)$$

where

$$L_d = L_{1\sigma} + k_{dm1} L_m \qquad (5.29)$$

$$L_m = \frac{6\mu_0 (k_{w1} N_1)^2 \tau l}{\pi^2 k_c (1 + k_{sd}) g p} \qquad (5.30)$$

and where $L_{1\sigma}$ is primary leakage inductance, L_m is magnetizing inductance, τ is pole pitch, l stack width, k_{w1} primary fundamental winding factor, k_c Carter coefficient, k_{sd} d-axis saturation factor, g = airgap, and p number of pole pairs.

The mechanical equations of motion are as follows:
Lateral motion (in the direction of thrust):

$$M\dot{u} = F_x - F_{xload} \qquad (5.31)$$

Normal motion (in the direction of normal force):

$$M\dot{u}_g = F_{na} - F_{nload} \qquad (5.32)$$

where u_g is the velocity in the normal direction. This last equation is considered in suspension control and in power loss calculations for normal motion. This power loss P_s is given by

$$P_s = F_{na} \dot{g} \qquad (5.33)$$

This loss can be taken into account in the actuator equivalent circuit by a resistor R_s given by

$$R_s = \frac{2(\omega_r L_d i_d)^2}{3 P_s} \tag{5.34}$$

Core losses may be represented by resistors R_{dm} and R_{qm} located across motion-induced voltages. If P_{diron} and P_{qiron} are the (measured or calculated) core losses for the fluxes in the two axes, then the respective values of the resistances are

$$R_{dm} = \frac{2(\omega_r L_q i_q)^2}{3 P_{diron}} \quad ; \quad R_{qm} = \frac{2(\omega_r L_d i_d)^2}{3 P_{qiron}} \tag{5.35}$$

Thus, we obtain the equivalent circuits shown in Figs. 5.11(a) and (b). Space-phasor diagrams corresponding to (5.17) through (5.19) and (5.22) may be drawn as shown in Figs. 5.12(a) and (b), where PMs are not present in Fig. 5.12(b). Clearly, PMs improve the power factor, as expected. Also, with PMs $F_x = 0$ for $i_d = 0$, but $F_x \neq 0$ for $i_q = 0$, whereas in the absence of PMs $F_x = 0$ for either $i_d = 0$ or $i_q = 0$.

5.4 STEADY-STATE CHARACTERISTICS

Because LRSAs are invariably used with power electronics control, their various characteristics are of interest only with vector control strategies [8], which are applied in conjunction with the dq-space-phasor model of the actuator. There are three main vector control strategies for LRSAs without PMs, as follows:

(i)   Constant i_d control
(ii) Constant i_d/i_q control
(iii) Constant primary flux λ_1 control

We now consider these strategies under steady-state operation of an LRSA. In secondary coordinates (with $d/dt = 0$) in dq-coordinates, (5.17) and (5.18) yield

$$V_d = R_1 I_d + \omega_r L_q I_q \tag{5.36}$$

$$V_q = R_1 I_q - \omega_r L_d I_d \tag{5.37}$$

and

$$V_1^2 = V_d^2 + V_q^2 \tag{5.38}$$

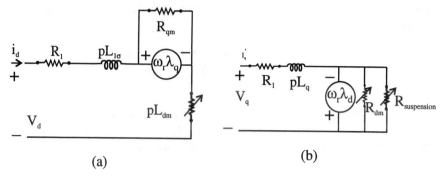

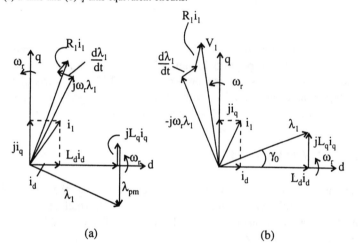

FIGURE 5.11

(a) *d*-axis and (b) *q*-axis equivalent circuits.

FIGURE 5.12

Space-phasor diagrams in secondary coordinates with (a) $\lambda_{PM} \neq 0$ and (b) $\lambda_{PM} = 0$.

Combining these equations with (5.21) gives

$$V_1^2 = (R_1^2 + \omega_r^2 L_d^2)I_d^2 + (R_1^2 + \omega_r^2 L_q^2)\left[\frac{2F_x\tau}{3\pi(L_d - L_q)I_d}\right]^2$$

$$- \frac{4R_1\omega_r F_x\tau}{3\pi} \tag{5.39}$$

Since $u = \tau\omega_r/\pi$, a dc series motor type of $F_x(u)$ curve is obtained from (5.39). The ideal no-load speed u_0 is given by

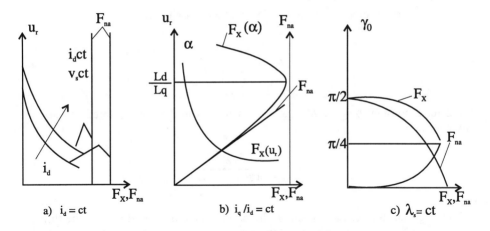

FIGURE 5.13
Thrust $[F_x(u)]$ and normal force $[F_{na}(u)]$ versus speed for (a) constant I_d; (b) constant I_d/I_q; and (c) constant primary flux λ_1.

$$u_0 = \frac{\tau \omega_{r0}}{\pi} = \frac{\tau \sqrt{V_1^2 - R_1^2 I_d^2}}{\pi L_d I_d} \tag{5.40}$$

Finally, with I_d and g constant, the normal force F_{na} is essentially constant. The various force characteristics are shown in Fig. 5.13. Because controlling I_d implies controlling the normal force, this type of control yields good results for a simultaneous propulsion and suspension control.

Next, we consider $i_d/i_q = \alpha = $ constant control. In this case the stator flux, under steady state, becomes

$$\lambda_1 = \sqrt{(L_d i_d)^2 + (L_q i_q)^2} = I_d \sqrt{L_d^2 + (\alpha L_q)^2} \tag{5.41}$$

The thrust can then be expressed as

$$F_x = \frac{3\pi}{2\tau}(L_d - L_q)\frac{\alpha \lambda_1^2}{L_d^2 + \alpha^2 L_q^2} \tag{5.42}$$

For maximum thrust, $\alpha = \alpha_k = L_d/L_q$. Thus, (5.42) yields

$$F_{xm} = \frac{3\pi}{4\tau}(L_d - L_q)\frac{\lambda_1^2}{L_d L_q} \tag{5.43}$$

The normal force F_{na}, from (5.28), can be approximated as

$$F_{na} = \frac{3}{2g} L_{dm} I_d^2 \qquad (5.44)$$

And for a constant α ($= L_d/L_q$) and a constant airgap the normal force can also be expressed as

$$F_{na} = \frac{\tau L_{dm} F_x}{\pi g (L_d - \alpha L_q)} \qquad (5.45)$$

indicating that, for constant α and g, the normal force is proportional to the thrust.
Combining (5.36) through (5.39) and denoting $L_d/L_q = \alpha$, we obtain

$$V_1^2 = \frac{2F_x \tau}{3\pi\alpha(L_d - L_q)} \left[(R_1 + \omega_r \alpha L_q)^2 + (\alpha R_1 - \omega_r L_d)^2\right] \qquad (5.46)$$

In this case a true dc series motor type of $F_x(u)$ curve is obtained for a given flux, as shown in Fig. 5.14(b). However, in this case the power factor is poorer. Also, F_x and F_{na} are not decoupled. So this type of control is applicable to thrust, or propulsion force, only.

Finally, the constant primary flux (λ_1 = constant) control may be applied with γ_0, the dq-flux angle, as a variable. This angle is defined by

$$\cos \gamma_0 = \frac{L_d I_d}{\lambda_1} \; ; \qquad \sin \gamma_0 = \frac{L_q I_q}{\lambda_1} \qquad (5.47)$$

The thrust F_x and the normal force F_{na} are given by

$$F_x = \frac{3\pi}{4\tau}(L_d - L_q) \frac{\lambda_1^2}{L_d L_q} \sin 2\gamma_0 \qquad (5.48)$$

$$F_{na} \simeq \frac{3}{2g} L_{dm} \frac{\lambda_1^2}{L_d^2} \cos^2 \gamma_0 \qquad (5.49)$$

The peak thrust is obtained at $\gamma_0 = \pi/4$, in which case (5.48) reduces to (5.43) for a given λ_1. As shown in Fig. 5.14(c), with controlled λ_1 and γ_0, F_x and F_{na} no longer depend on speed, and independent propulsion and suspension control may be exer-

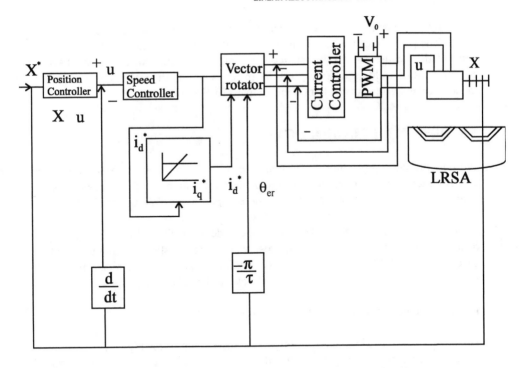

FIGURE 5.14
Position vector current control of an LRSA (with I_d = constant or I_q/I_d = constant).

cised. Note that for stability $\gamma_0 \leq \pi/4$. This type of control is effective in suspension control since L_{dm} and L_d in (5.49) depend only on the airgap.

For I_q/I_d = constant and λ_1 = constant, the ideal power factor angle ϕ_{1k} is given by

$$\tan \phi_{1k} = \frac{1 + L_q/L_d}{1 - L_q/L_d} \tag{5.50}$$

indicating that $\phi_{1k} > \pi/4$ and the power factor cannot exceed 0.707. Copper losses do increase the power factor to some degree.

5.5 VECTOR CONTROL ASPECTS

For a combined propulsion and suspension control the direct flux and thrust controller, presented in Chapter 3 for LIAs, may be applied, with some modifications, to LRSAs. For thrust-only control, I_d = constant and I_d/I_q = constant vector control strategies may be applied, as is done for rotary reluctance synchronous motors [9,10]. A block diagram of such a control system is shown in Fig. 5.14. A linear position sensor provides for position feedback for the speed controller and for estimating θ_{er} for the vector rotator. Position, speed, and current controllers produce the required positioning target. The power electronics consisting of a PWM voltage-source inverter is current controlled. Because the maximum frequency of the fundamental is below 60 Hz, ac current control yields satisfactory results in terms of current response. In summary, most ac motor vector controls may be adapted LRSAs [8].

5.6 DESIGN METHODOLOGY

We now present a procedure for designing an LRSA. Consider an application where constant rated thrust is required from standstill to maximum speed. A flat single-sided actuator with a moving primary is considered appropriate. We use the peak (short-duration) thrust ($\gamma_0 = \pi/4$) criterion to reduce the mass of the primary. The actuator is fed from a voltage-source PWM inverter, with a dc link voltage $V_0 = 500$ V dc. Other data are as follows: $F_{xmax} = 2$ kN (for a short duration) at standstill and up to a base speed u_b; $F_x = 0.8$ kN for continuous duty; maximum speed $u_{max} = 2$ m/s; travel length $l_t = 2$ m; airgap $g = 0.5$ mm. We proceed now in the following sequence.

Primary Geometry and Maximum Slot mmf

For $g = 0.5$ mm, a pole pitch $\tau = 200g = 0.1$ m is required for high L_{dm}/L_{qm} ratio for a high performance. From Figs. 5.10 and 5.11 we obtain $L_{dm}/L_{qm} = 20$ and $L_{dm} = 0.97$ L_m for $\tau_p/\tau = 2/3$. Assuming $L_{s\sigma} \approx 0.06\ L_m$ and a d-axis saturation factor $k_{sd} = 0.3$ we have

$$\frac{L_d}{L_q} = \frac{0.97 L_m/(1 + k_{sd}) + L_{s\sigma}}{0.97 L_m/20 + L_{s\sigma}} = 7.43 \tag{5.51}$$

At a constant flux, the maximum thrust recurs at $\gamma_0 = \pi/4$ and $I_q/I_d = L_d/L_q$. Now, if B_{d1p} is the airgap flux density produced by I_d and B_{g1k} is the resultant airgap density for maximum thrust, we get

$$B_{d1p} = \frac{B_{g1k}}{\sqrt{1 + (L_d/L_q)^2 \left[L_{qm}(1 + k_{sd})^2/L_{dm}\right]^2}} \tag{5.52}$$

Assuming $B_{g1k} = 0.75$ T and substituting other numerical values in (5.52) yield

$$B_{d1p} = \frac{0.75}{\sqrt{1 + 7.43^2(1.3/20)^2}} = 0.675 \text{ T} \tag{5.53}$$

This flux density is related to the primary mmf by

$$B_{d1p} = \frac{3\sqrt{2}}{\pi k_c(1 + k_{sd})gp} \mu_0(k_{w1}N_1)k_{dm1}I_{dp}$$ (5.54)

where k_{w1} is the primary winding factor, N_1 the primary number of turns, k_{sd} the d-axis saturation factor, k_c the Carter coefficient, g the airgap, p the number of pole pairs, and I_{dp} the d-axis component of the phase current.

Now, if we choose $q = 2$ and coil span/pole-pitch ratio $y/\tau = 5/6$, $k_{w1} = 0.933$. Thus, (5.54) gives

$$N_1I_{dp} = \frac{0.675 \times \pi \times 1.3 \times 1.3 \times 0.5 \times 10^{-3}p}{3\sqrt{2} \times 4\pi \times 10^{-7} \times 0.933} = 360p \text{ At}$$ (5.55)

which is the d-axis mmf. The q-axis mmf is given by

$$N_1I_{gp} = N_1I_{dp}\left(\frac{L_d}{L_q}\right) = 360p \times 7.43 = 2675p \text{ At}$$ (5.56)

In terms of the d- and q-axis phase currents, I_{dp} and I_{qp} respectively, the thrust may be expressed as

$$F_{xp} = (\sqrt{3})^2\frac{\pi}{\tau}L_m\left(\frac{k_{dm1}}{1 + k_{sd}} - \frac{k_{qm1}}{k_{dm1}}\right)I_{dp}I_{qp}$$ (5.57)

Or substituting for L_m in (5.57) yields

$$F_{xp} = \frac{3\pi}{\tau}\left[\frac{6\mu_0(k_{w1}N_1)^2\tau l}{\pi^2 k_c gp}\right]\left(\frac{k_{dm1}}{1 + k_{sd}} - \frac{k_{qm1}}{k_{dm1}}\right)I_{dp}I_{qp}$$ (5.58)

With the given numerical values, we obtain from (5.58)

$$2000 = \frac{18 \times 4\pi \times 10^{-7} \times 0.933^2 \times 360p \times 2675pl}{\pi \times 1.3 \times 0.5 \times 10^{-3} \ p}\left(\frac{0.97}{1.3} - \frac{1}{20}\right)$$

or

$$pl = 0.309 \text{ m}$$ (5.59)

With $q = 2$, the slot peak mmf is given by

$$n_s I_p = \frac{N_1 I_p}{pq} = \frac{360p \sqrt{1 + 7.43^2}}{p \times 2} = 1350 \text{ At/slot} \tag{5.60}$$

We choose $p = 3$-pole pairs and the stack width $l = 0.102$ m. Thus, the peak thrust density is

$$f_{xp} = \frac{F_{xp}}{2p\tau l} = \frac{2000}{2 \times 3 \times 0.103 \times 0.1 \times 10^4} = 3.24 \text{ N/cm}^2 \tag{5.61}$$

which is a reasonable value.

Primary Slot Geometry

For a 25% duty cycle, the average slot mmf is

$$n_s I_{av} = n_s I_p \sqrt{0.25} = 1350\sqrt{0.25} = 675 \text{ At/slot} \tag{5.62}$$

The slot active area is given by

$$A_{slot} = \frac{n_s I_{av}}{k_{fill} J_{av}} \tag{5.63}$$

If we choose k_{fill} = fill factor = 0.44 and J_{av} = average current density = 4.5 A/mm², (5.62) and (5.63) yield

$$A_{slot} = \frac{675}{0.44 \times 4.5} = 341 \text{ mm}^2 \tag{5.64}$$

Since $\tau = 0.1$ m and $q = 2$, the slot pitch τ_s is

$$\tau_s = \frac{100}{2 \times 3} = 16.67 \text{ mm} \tag{5.65}$$

For $B_{g1k} = 0.75$ T, the slot width may be taken as $b_s = 8$ mm. Thus, slot depth $h_s = 341/8 = 42.7$ mm, and we obtain the slot geometry shown in Fig. 5.15. The stack core thickness h_{cp} (Fig. 5.15) is obtained from

$$h_{cp} \approx \frac{B_{g1p}}{B_{cp}}\left(\frac{\tau}{\pi}\right) = \frac{0.75}{1.3} \times \frac{0.1}{\pi} = 18.36 \text{ mm} \approx 20 \text{ mm} \tag{5.66}$$

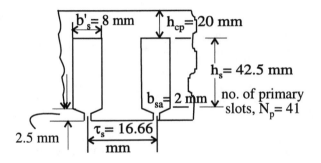

FIGURE 5.15
Primary slot geometry.

Finally, the primary core length l_p is given by

$$l_p = (2p + 1)qm\tau_s = (2 \times 3 + 1)2 \times 3 \times 16.67 \approx 700 \text{ mm} \qquad (5.67)$$

where m, the number of phases, is 3, and q, the number of slots/pole per phase, is 2, or the total number of slots $N_p = (2p + 1)qm / 1 = (2 \times 3 + 1)2 \times 3 / 1 = 41$ slots.

Electrical Parameters

For steady-state analysis of the LRSA the important parameters to be evaluated are the primary resistance R_1, and the d- and q-axis inductances L_d and L_q, respectively. These are determined as follows:

The primary resistance is given by

$$R_1 = \frac{\rho_c^2(l + 0.01 + 1.5y)N_1^2}{N_1 I_{av}/J_{av}} \qquad (5.68)$$

where $\rho_c = 2.3 \times 10^{-8}$ Ω-m is the resistivity of copper and other symbols have been defined earlier. Substituting numerical values in (5.68) gives

$$R_1 = \frac{2.3 \times 10^{-8} \times 2(0.1 + 0.01 + 1.5 \times 0.083)N_1^2}{6 \times 675/4.5 \times 10^6} = 1.2 \times 10^{-5}N_1^2 \qquad (5.69)$$

The formula for the magnetizing inductance is

$$L_m = \frac{6\mu_0(k_{w1}N_1)^2 l\tau}{\pi^2 pk_c g} \qquad (5.70)$$

which yields

$$L_m = \frac{6 \times 4\pi \times 10^{-7}(0.933N_1)^2 \times 0.102 \times 0.1}{\pi^2 \times 3 \times 1.3 \times 0.5 \times 10^{-3}} = 3.48 \times 10^{-6}N_1^2 \quad (5.71)$$

With $k_{sd} = 0.3$, $k_{dm1} = 0.97$, and $k_{qm1} = k_{dm1}/20$, we have

$$L_{dm} = \frac{k_{dm1}}{1 + k_{sd}} L_m = \frac{0.97}{1.3} \times 3.48 \times 10^{-6} \, N_1^2 = 2.6 \times 10^{-6} \, N_1^2 \quad (5.72)$$

$$L_{qm} = \frac{k_{qm1}}{k_{dm1}} L_m = \frac{1}{20} \times 3.48 \times 10^{-6}N_1^2 = 0.174 \times 10^{-6}N_1^2 \quad (5.73)$$

The leakage inductance $L_{1\sigma}$, is obtained from

$$L_{1\sigma} = \frac{2\mu_0}{pq}(\lambda_s l + \lambda_e l_{ec})N_1^2 \quad (5.74)$$

where

$$\lambda_s = \text{slot permeance} = \frac{h_s}{3b_s} + \frac{h_{sa}}{b_{sa}} \quad \text{(see Fig. 5.15)} \quad (5.75)$$

• $$\lambda_e = 0.3q(3\beta_1 - 1), \quad \beta = \text{chording factor} = 5/6 \quad (5.76)$$

and l_{ec} = length of end connection. Thus, (5.75) and (5.76) yield

$$\lambda_s = \frac{42.7}{3 \times 8} + \frac{2}{8} = 2.03 \quad (5.77)$$

$$\lambda_e = 0.3 \times 2\left(\frac{3 \times 5}{6} - 1\right) = 0.9 \quad (5.78)$$

Consequently, (5.74), (5.77), and (5.78) give

$$L_{1\sigma} = \frac{2 \times 4\pi \times 10^{-7}}{3 \times 2} \left[2.03 \times 0.1 + 0.9\,(0.01 + 1.5 \times 0.083)\right]N_1^2$$

$$= 1.35 \times 10^{-7} \, N_1^2 \quad (5.79)$$

Finally,

$$L_d = L_{1\sigma} + L_{dm} = (1.35 + 26) \times 10^{-7} N_1^2 = 2.735 \times 10^{-6} N_1^2 \qquad (5.80)$$

$$L_q = L_{1\sigma} + L_{qm} = (1.35 + 1.74) \times 10^{-7} N_1^2 = 3.09 \times 10^{-7} N_1^2 \qquad (5.81)$$

The ratio $L_d/L_q = 27.35/3.09 = 8.8$ and *not* 7.43 as originally assumed in (5.51). Thus, the designed actuator is capable of supplying about 10% additional thrust.

Number of Turns per Phase

The number of turns per phase, N_1, must be adequate to provide for the rated thrust at rated (maximum) speed and yield sufficient voltage for maximum thrust at standstill and up to the base speed u_b. The rms value of the ceiling phase voltage is given by

$$V_p = \frac{4k_{v0}V_0}{\sqrt{6}\,\pi} = \frac{4 \times 0.85 \times 500}{\sqrt{6}\,\pi} = 220 \text{ V} \qquad (5.82)$$

The voltage for the *dq*-model is

$$V_1 = \sqrt{2}\,V_p = 220\,\sqrt{2} = 311 \text{ V} \qquad (5.83)$$

Assuming that the rated thrust is obtained at maximum power factor, we have

$$\frac{I_q}{I_d} = \sqrt{\frac{L_d}{L_q}} = \sqrt{8.8} = 2.97 \qquad (5.84)$$

and

$$\frac{F_x}{F_{xp}} = \left(\frac{N_1 I_d}{N_1 I_{dp}}\right)^2 \times \frac{2.97}{7.43}$$

Thus,

$$N_1 I_d = N_1 I_{dp} \sqrt{\frac{800 \times 7.43}{2000 \times 2.97}} = N_1 I_{dp} \times 1.0 = 360 \times 3 \times 1.0 = 1080 \text{ At} \qquad (5.85)$$

$$N_1 I_q = 2.97 \times N_1 I_d = 2.97 \times 1080 = 3207 \text{ At} \qquad (5.86)$$

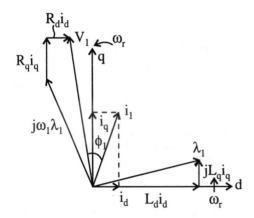

FIGURE 5.16
Phasor diagram of an LRSA.

Now, in order to relate the voltages and currents to the electrical parameters, we refer to the phasor diagram of Fig. 5.16, from which

$$V_d = R_1 I_d - \omega_r L_q I_q \qquad (5.87)$$

$$V_q = R_1 I_q + \omega_r L_d I_d \qquad (5.88)$$

Substituting $\omega_r = u\pi/\tau = 2\pi/0.1 = 62.8$ r/s and other numerical values in (5.87) and (5.88) and combining with (5.85) and (5.86) we obtain

$$V_d = 1.2 \times 10^{-5} N_1^2 I_d - 62.8 \times 3.09 \times 10^{-7} N_1^2 I_q$$

$$= (1.2 \times 10^{-5} \times 1080 - 194 \times 10^{-7} \times 3207) N_1 = -0.049 N_1 \qquad (5.89)$$

$$V_q = 1.2 \times 10^{-5} N_1^2 I_q + 62.8 \times 2.735 \times 10^{-6} N_1^2 I_d$$

$$= (1.2 \times 10^{-5} \times 3207 + 171.76 \times 10^{-6} \times 1080) N_1 = 0.224 N_1 \qquad (5.90)$$

Since

$$V_p = \sqrt{V_d^2 + V_q^2} \qquad (5.91)$$

we have

$$V_p = 220 = \sqrt{(-0.049)^2 + (0.224)^2} \; N_1 = 0.23 N_1 \qquad (5.92)$$

Finally, $N_1 = 220/0.23 \approx 960$ turns. The number of turns per coil is given by

$$N_1 = 2pqn_c \tag{5.93}$$

or, $n_c = 960/2 \times 3 \times 2 = 80$.

Performance Calculations

The currents I_d and I_q are obtained from

$$I_d = \frac{N_1 I_d}{N_1} = \frac{1080}{960} = 1.13 \text{ A} \tag{5.94}$$

$$I_q = \frac{N_1 I_q}{N_1} = \frac{3027}{960} = 3.15 \text{ A} \tag{5.95}$$

The phase current is given by

$$I_p = \sqrt{I_d^2 + I_q^2} = \sqrt{1.13^2 + 3.15^2} = 3.35 \text{ A} \tag{5.96}$$

As a measure of performance, we determine the product of coefficiency η and the power factor $\cos \phi_1$ from

$$\eta \cos \phi_1 = \frac{F_x u_m}{3 V_p I_p} = \frac{800 \times 2}{3 \times 220 \times 3.35} = 0.72 \tag{5.97}$$

From (5.68), (5.80), and (5.81) the electrical parameters are $R_1 = 1.2 \times 10^{/5} \times 960^2 = 11 \ \Omega$; $L_d = 2.735 \times 10^{/6} \times 960^2 = 2.52$ H; and $L_q = 3.09 \times 10^7 \times 960^2 = 0.285$H. Thus,

Copper losses $= 3I_p^2 R_1 = 3 \times 3.35^2 \times 11 = 370$ W
Output power $= 800 \times 2 = 1600$ W
Electrical efficiency $= 1600/(1600 + 370) = 0.81$
Power factor $= 0.72/0.81 = 0.89$

At standstill, $\omega_r = 0$. So (5.56) and (5.87) yield $V_{dp} = R_1 I_{dp} = 11 \times 1.13 = 12.43$ V, and $V_{qp} = 12.43 \times 7.43 = 92.35$ V. Thus, $V_p = (V_{dp}^2 + V_{qp}^2)^{\frac{1}{2}} = (12.43^2 + 92.35^2)^{\frac{1}{2}} = 93$ V. With these phase voltages, in order to maintain the peak thrust up to a base speed ω_b and with a 220 V per phase ceiling we have

$$V_{dp} = 12.43 - \omega_b \times 3.09 \times 10^{-7} \times 960 \times 1080 \times 7.43$$

$$= 12.43 - 2.38 \omega_b \tag{5.98}$$

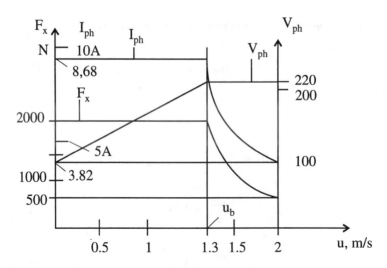

FIGURE 5.17
Thrust/speed and current/speed envelopes.

$$V_{qp} = 93 + \omega_b \times 2.735 \times 10^{-6} \times 960 \times 1080 = 93 + 2.83\,\omega_b \qquad (5.99)$$

Consequently,

$$V_{dp}^2 + V_{qp}^2 = 220^2 = (12.43 - 2.38\,\omega_b)^2 + (93 + 2.83\,\omega_b)^2 \qquad (5.100)$$

Solving for ω_b from (5.100) yields $\omega_b \approx 40$ r/s. This value corresponds to a base speed of $u_b = \tau\omega_b/\pi = 0.1 \times 40/\pi \approx 1.3$ m/s.

The maximum rms phase current, $\quad I_{phm} = \dfrac{n_s I_p pq}{N_1} = \dfrac{1350 \times 3 \times 2}{960} = 8.43$ A

Copper loss at peak current, $\quad P_{co} = 3 \times 8.43^2 \times 11 = 2349$ W
Efficiency at peak thrust for base speed,

$$\eta_m = \frac{F_{xp}u_b}{F_{xp}u_b + P_{cop}} = \frac{2000 \times 1.3}{2000 \times 1.3 + 2349} = 0.525$$

Power factor at peak thrust,

$$\cos \phi_p = \frac{F_{xp}u_b}{3V_p I_{phm}\eta_m} = \frac{2000 \times 1.3}{3 \times 220 \times 8.43 \times 0.525} = 0.89$$

The preceding results are graphically represented in Fig. 5.17.

Mass of the Primary

The primary mass is composed of the following: mass of the winding, G_{co}; mass of teeth, G_t; and mass of the core (or back iron), G_{core}. These are calculated as follows:

$$G_{co} = 3N_1 \frac{I_{av}}{J_{av}} \times 2(l + 0.01 + 1.5 \times \frac{5}{6} \times 0.1) \times \text{density of copper}$$

$$= \frac{3 \times 6 \times 675}{4.5 \times 10^6} \times 2 \left(0.1 + 0.001 + \frac{0.75}{6}\right) \times 8900$$

$$= 11.3 \text{ kg}$$

$$G_t = 43 \, b_t h_t l \times \text{density of iron}$$

$$= 43 \times 8 \times 45 \times 0.1 \times 10^{-6} \times 7900 = 12.23 \text{ kg}$$

$$G_{core} = 42 \tau_s l h_{cp} \times \text{density of iron}$$

$$= 42 \times 16.67 \times 10^{-3} \times 0.1 \times 0.02 \times 7900 = 11.07 \text{ kg}$$

Total mass of the primary $= 11.3 + 12.33 + 11.07 = 34.6$ kg

Peak thrust/weight $= \dfrac{2000}{34.6} = 57.8$ N/kg

The above calculated values compare favorably with those of LIAs.

REFERENCES

1. B. J. Chalmers and A. C. Williamson, *AC machines: electromagnetics and design.* (Wiley, New York, 1991), pp. 75-85.
2. P. J. Lawrenson and L. A. Agu, "Theory and performance of polyphase reluctance machines," *Proc. IEE*, vol. 111, 1964, pp. 1425-1435.
3. S. A. Nasar and I. Boldea, *Linear motion electric machines* (Wiley-Interscience, New York, 1976), Chapter 6.
4. D. A. Staton, T. J. E. Miller, and S. E. Wood, "Maximizing the saliency ratio of the synchronous reluctance motor," *Proc. IEE*, vol. 140(B), 1993, pp. 249-259.
5. I. Boldea et al., "Distributed anisotropy rotor synchronous (DARSYN) drives—motor identification and performance," *Rec. ICEM*, 1992, vol. 2, pp. 542-548.
6. R. Mayer et al., "Inverter-fed multiphase reluctance machine with reduced armature reaction and high power density," *Rec. ICEM*, 1986, pt. III, pp. 1138-1141.
7. I. Boldea, Z. Fu, and S. A. Nasar, "Performance evaluation of ALA rotor synchronous motors," *IEEE Trans.*, vol. IA-30, 1994, pp. 977-985.
8. I. Boldea and S. A. Nasar, *Vector control of ac drives* (CRC Press, Boca Raton, FL, 1992).

9. I. Boldea, Z. Fu, and S. A. Nasar, "Digital simulation of vector current controlled axially laminated anisotropic (ALA) rotor synchronous motor servo drive," *Elec. Mach. Power Syst.*, vol. 19, 1991, pp. 415-424.

10. L. Xu et al., "Vector control of a synchronous reluctance motor, including saturation and iron losses," *Rec. IEEE-IAS*, 1990 Annual Meeting, 1990 vol. I, pp. 359-364.

CHAPTER

6

LINEAR SWITCHED RELUCTANCE ACTUATORS

Linear switched reluctance actuators (LSRAs) are counterparts of rotary switched reluctance motors (SRMs). Except for the determination of normal forces, the theory developed for SRMs is almost entirely applicable to LSRAs [1]. Just like the SRM, the LSRA produces a thrust and motion by the tendency of a ferromagnetic secondary to assume a position where the inductance of the excited primary is maximized. In other words, the LSRA is an electromagnetic force device. However, it is a multiphase device, with only one phase energized, generally. The turning on and off of each phase is triggered by a linear position sensor or an estimator. LSRAs are a unique choice where reliability is a major concern, because they are capable of operating even if one phase gets shorted or disconnected.

6.1 PRACTICAL CONFIGURATIONS

Relatively speaking, the literature on LSRAs is sparse [2-5]. However, a few pages are available on flat and tubular configurations of LSRAs. A primitive single-sided geometry is shown in Fig. 6.1, in which either the primary or the secondary could be the mover. Notice that an LSRA is a doubly salient device. The primary and the secondary poles (or teeth) have about the same width, $w \simeq b_s$, but the respective slot pitches τ_p and τ_s are related by

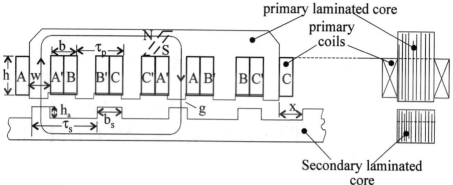

FIGURE 6.1
Single-sided flat three-phase LSRA.

$$2\tau_s = m\tau_p \tag{6.1}$$

where m is the number of phases. Also, the primary slot width b should be slightly greater than the secondary tooth width b_s in order to have a low minimum inductance of the primary phases. For the relative primary and secondary positions shown in Fig. 6.1, phase A is with both coils in the maximum inductance, or zero thrust, position; that is, phase A has just terminated its conduction period for actuating. Either phase B or C is next turned on for the motion of the secondary to the left or right (Fig. 6.1). The three phases share the back cores of the primary and the secondary but the teeth (or poles) experience the flux corresponding to the phase which is conducting. For a large τ_p, the flux lines in the cores are long.

Thrust or force F_e is produced in an LSRA by virtue of the variation of magnetic coenergy W_m' stored in the airgap with the secondary position for a constant primary current; that is,

$$F_e = \left.\frac{\partial W_m'}{\partial x}\right|_{i = const} \tag{6.2}$$

In the absence of saturation (6.2) simplifies to

$$F_e = \frac{1}{2}i^2\frac{\partial L}{\partial x} \tag{6.3}$$

The stroke length is practically equal to b_s or $\tau/3$. Thrust and normal forces occur between primary and secondary teeth pertaining to the conducting phase. Because of sequential switching of the phases, noise and vibration are likely to occur, especially

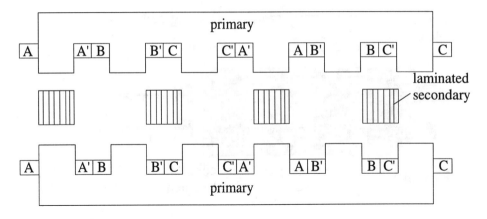

FIGURE 6.2
Double-sided flat three-phase LSRA.

since the normal force is much greater than the thrust. A double-sided LSRA ideally does not have a net normal force exerted on the secondary. Nevertheless, the normal force on each side of the stator will be high, thereby creating stresses on the stator frame [4]. A double-sided SRLA may have multiple airgaps, as shown in Fig. 6.2. Such a configuration has a lower minimum inductance, because for an unaligned position (for phases B and C in Fig. 6.2) there are four reluctances in series.

Whereas up to this point we have discussed LRSAs having transverse fields, an LSRA with a longitudinal field is shown in Fig. 6.3. In this case, the coils are ring shaped and the positions of minimum and maximum inductances are interchanged, as compared with a transverse field LSRA [5]. It has been claimed that a longitudinal field LSRA has greater force densities.

A tubular configuration suitable for short travel (0.4 to 0.5 m) and high thrust density applications is shown in Fig. 6.4. It uses longitudinal lamination stacks on the primary and on the secondary, with ring-shaped coils on the primary. The secondary has two sets of laminations of different outer diameters. The tubular LSRA makes a good use of copper as end connections are almost nonexistent.

Completely separate magnetic circuits, or cores, for each phase may be configured for each phase to reduce the length of the flux path in the back iron, and thereby lower the magnetizing mmf, especially when the core saturates. But the utilization of the core material is poor, as the back iron is not shared by more than one phase. For some applications where high temperatures do not exist and reliability is important, permanent magnets (PMs) may be added on either the primary or the secondary. Such magnets are shown on the primary of the actuator in Figs. 6.1 and 6.4. In such a case, the PM flux λ_{PM} interacts with the coil flux to produce the thrust. The PM may be made to increase or decrease the net thrust, where the corresponding thrust is given by

$$F_{xPM} = \frac{\partial \lambda_{PM}}{\partial x} i \qquad (6.4)$$

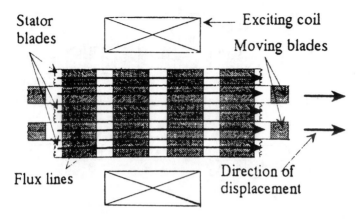

FIGURE 6.3
Tubular multi-airgap LSRA [5].

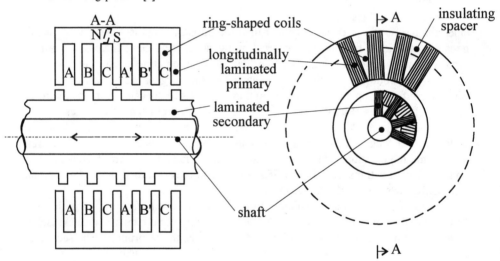

FIGURE 6.4
Tubular LSRA.

PMs tend to reduce the machine inductances and the static power ratings.

As with other reluctance-type actuators, the normal force (which is attraction type) in an LSRA is given by the general form

$$F_{na} = \left. \frac{\partial W'_m}{\partial g} \right|_{i=const} \tag{6.5}$$

In the absence of saturation (6.6) reduces to

$$F_{na} = \frac{1}{2} i^2 \left. \frac{\partial L}{\partial g} \right|_{i=const} \tag{6.6}$$

where g is some airgap.

Finally, as a guideline, the number of phases and the ratio between the primary and secondary slot pitches may be taken as that in a rotary switched reluctance motor [1].

6.2 INSTANTANEOUS THRUST

Neglecting saturation, from (6.3) it follows that a thrust is developed as long as $\partial L/\partial x \neq 0$. Also, if L increases with x ($\partial L/\partial x$ positive), an actuating force is obtained, whereas for a negative $\partial L/\partial x$ a regenerative (braking) force is produced. This latter force acts in a direction opposite to that of the desired motion. Ideal variations of a phase inductance and actuating and braking forces are shown in Fig. 6.5, where the LSRA current control with position is provided by a two-quadrant dc-dc converter per phase. As shown in Fig. 6.5, as the current increases, the inductance decreases drastically. Thus, the ratio of unsaturated aligned (L_{au}) to unaligned (L_u) inductance, L_{au}/L_u, decreases from about 10 to 15 to almost 7 or 8. If the current in a phase is constant and the flux varies linearly with the position, then the thrust remains constant with the mover position. The thrust developed by an unsaturated three-phase LSRA, supplied with currents having waveforms shown in Fig. 6.6(a), is pulsating during the overlapping of two phases [Fig. 6.6(b)].

As evident from Figs. 6.5 and 6.6, the current in a high-inductance, close-to-aligned position is turned off for actuating mode and turned on for regenerative braking. A basic two-quadrant dc-dc converter (Fig. 6.7) is required for each phase to turn it on or off. When the transistor T_1 conducts, the current increases until the motion-induced voltage E in the phase equals the dc bus voltage V_0 less the resistive drop; that is,

$$E = V_0 - I_b R_1 \tag{6.7}$$

where

$$E = \left(\frac{\partial \lambda}{\partial x} \right) u_b \tag{6.8}$$

u_b is base linear speed, and I_b base current. If the flux varies linearly with position, then

$$\frac{\partial \lambda}{\partial x} = C_\lambda(i) \tag{6.9}$$

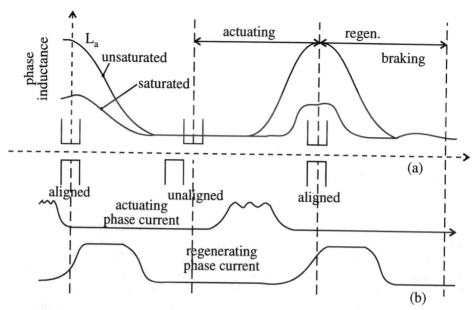

FIGURE 6.5
Phase inductance and phase current versus position.

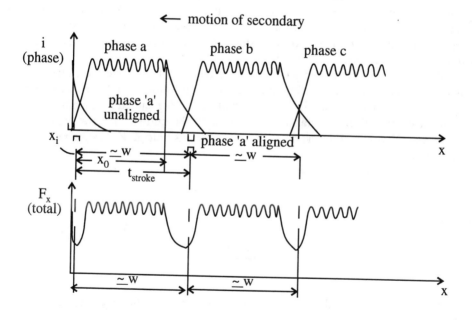

FIGURE 6.6
Total thrust versus position for flat-top current and ideal (linear) phase flux/position increase.

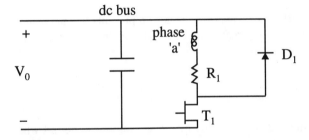

FIGURE 6.7
Basic two-quadrant dc-dc converter for one phase.

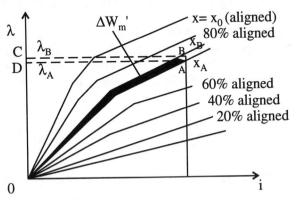

FIGURE 6.8
λ-i curves for different positions x.

Consequently, the base speed for a given base current may be obtained iteratively from (6.7) through (6.9) as $C_\lambda(i)$ is not a linear function in the presence of saturation. In practice, C_λ is a function of position also, and decreases with position from turn-on position x_i to turn-off position x_0. If the speed is below u_b, the current may be chopped and maintained constant even at a higher value current than I_b.

To determine the thrust in the presence of saturation, $\lambda(i)$ curves for given values of position, from x_i to x_0, are required. These curves can be obtained from field plots by the finite-element method (FEM). Qualitatively, a family of $\lambda(i)$ curves are shown in Fig. 6.8. Having obtained $\lambda(i)$ we determine the stored magnetic coenergy $W_m\prime$ from

$$W_m' = \int \lambda \, di \qquad (6.10)$$

The thrust is then obtained from (6.2). For two positions x_A and x_B close to each other, (6.2) may also be written as

$$F_e(x) = \frac{\Delta W_m'}{x_B - x_A} \qquad (6.11)$$

where $\Delta W_m'$ is the area OAB in Fig. 6.8. The area OAB also represents the mechanical work done. As expected, the developed thrust is not constant, and depends on position and level of saturation.

6.3 AVERAGE THRUST

The average thrust can be computed from the energy cycle. During the period when the transistor conducts (OBC in Fig. 6.9), the current decreases from i_B to i_C (or B to C in Fig. 6.9) toward the end of the stroke if the motion-induced voltage increases to a point that I_h cannot be maintained. From O to C, on the trajectory OBC, the $\lambda(i)$ curves jump with a change in position from x_i to x_0. Notice that $x_0 < w$ and the turn-off process begins before the flux reaches its maximum value. This process is along CDO. The total energy converted to mechanical work per phase per stroke is W_m. The energy returned to the dc source during current turn-off (D_1 conducting, Fig. 6.7) by discharging the energy stored in phase coil is R. Thus, we may define an energy conversion ratio

$$\eta_E = \frac{W_m}{W_m + R} \tag{6.12}$$

The total average thrust F_{xav} for m phases for a constant speed is determined from

$$F_{xav} = \frac{mW_m}{l_{stroke}} \; ; \quad l_{stroke} = w \tag{6.13}$$

Thrust pulsations occur if the current is constant and the core is saturated. It may be possible to reduce the thrust pulsations by shaping the current waveforms.

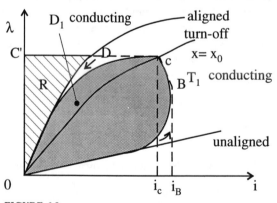

FIGURE 6.9
Energy conversion cycle for average thrust.

6.4 CONVERTER RATING

To power an LSRA, the static power converter must be able to handle the total energy $(W_m + R)$. The peak flux linkage per phase λ_c during a conducting period t_c at a constant speed u (with the phase resistance $R_1 \approx 0$) is related to the dc bus voltage V_0 by

$$\lambda_c \approx V_0 t_c = \frac{V_0}{u} (x_0 - x_i) \tag{6.14}$$

In terms of energy, we have

$$W_m + R = \frac{W}{\eta_E} = k\lambda_c I_c = \frac{kV_0}{u} (x_0 - x_i)I_c \tag{6.15}$$

where I_c is the peak current in the controller and k is a constant known as the converter utilization factor [1]. Consequently, the peak volt-amp rating of an m-phase converter becomes

$$S_{peak} = mV_0 I_c = \frac{F_{xav} l_{stroke} u}{k\eta_E(x_0 - x_i)} \tag{6.16}$$

We can define electromagnetic power P_{el} by

$$P_{el} = F_{xav} u \tag{6.17}$$

Thus, (6.16) and (6.17) can be combined to obtain

$$S_{peak} = \frac{P_{el} l_{stroke}}{k\eta_E(x_0 - x_i)} \tag{6.18}$$

Now, the magnetization curves depend on the ratio of the pole width to the airgap, w/g, and on the secondary height h_s. Therefore, η_E and k both vary with w/g and h_s/g. Also, because the turn-off position x_i occurs before the end of travel, l_{stroke}, in practice S_{peak} must be greater than that given by (6.18). For numerical values $\eta_E < 0.65$, $k < 0.7$ to 0.8, and $l_{stroke}/(x_0 - x_1) \approx 0.8 S_{peak}/P_{el} \approx 10$ kVA/kW.

6.5 STATE-SPACE EQUATIONS

To investigate the dynamics of an LSRA, we develop the actuator and converter equations while one phase is conducting. With one transistor per phase (Fig. 6.7) the state equations are

$$\frac{d\lambda(i,x)}{dt} = V_0 - R_1 i - V_{T1}, \qquad x_i < x < x_0 \tag{6.19}$$

$$T_1 \text{ conducting}$$

$$\frac{d\lambda(i,x)}{dt} = -V_0 - Ri + V_{D1}, \qquad x_0 < x < x_i + l_{stroke} \tag{6.20}$$

$$D_1 \text{ conducting}$$

$$M \frac{du}{dt} = F_x - F_{load} \tag{6.21}$$

$$\frac{dx}{dt} = u \tag{6.22}$$

where $F_x = \partial W_m'/\partial x = $ thrust, V_{T1} is the transistor voltage drop, and V_{D1} the diode voltage drop. These equations are valid for x ranging from x_i to $(x_i + l_{stroke})$, which corresponds to the period when phase a is conducting. Phase b or c is turned on at $x_0 < x_i + l_{stroke}$.

Once the family of function $\lambda(i,x)$ is known, the family of coenergy function $W_m'(i,x)$ curves is obtained. Approximate analytical expressions for $\lambda(i,x)$ and $W_m'(i,x)$ are then found. These expressions must have continuous derivatives with respect to x, u, and x as variables. The variable i is eliminated by calculating it at each integration step for given x and λ.

6.6 CONTROL ASPECTS

LSRAs can be used where thrust and/or linear positioning control is required. Positioning control implies thrust or current control. A refined linear position sensor is required for position feedback when precise positioning is desired. A block diagram showing a control system for an LRSA is shown in Fig. 6.10 [3]. The position sensor also determines the turn-on and turn-off positions for each phase. This task may be done with a three-element Hall-type single-position sensor, which does not, however, allow for advancing the turn-on or turn-off positions in each phase if desired for a precise thrust control. The amplitude of the current may be kept constant or may be varied with position during the conduction period to smooth out the thrust pulsations. A sliding mode controller may be used to deal with the severe nonlinearities of the actuator. A first-order functional S is adequate such that

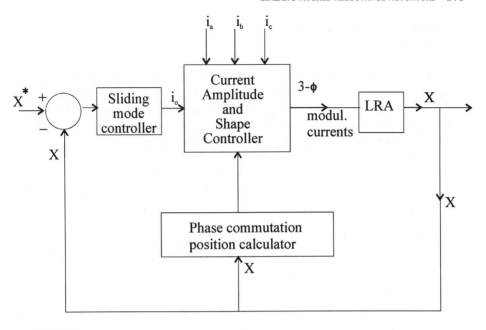

FIGURE 6.10
Positioning control system.

$$S = (x^* - x) - T_s \frac{dx}{dt} \tag{6.23}$$

with the control law:

$$
\begin{aligned}
i_0^* &= +I_{max} \quad \text{for} \quad S > 0 \\
&= -I_{max} \quad \text{for} \quad S < 0
\end{aligned}
\tag{6.24}
$$

The functional S can be scanned at the rated switching frequency in the converter. In this case, current amplitude controllers are used for current protection.

6.7 DESIGN METHODOLOGY

LSRAs are highly nonlinear devices, especially owing to saturation. Saturation limits the maximum flux λ_c in the actuator and thereby limits the dc link voltage for a given speed, as governed by (6.14). In the absence of a standard design procedure, we design an LSRA using the following data:

Rated thrust up to base speed, $F_{xr} = 500$ N
Base speed, $u_b = 2$ m/s
dc link voltage, $V_0 = 300$ V dc

Number of phases, $m = 3$
Topology, tubular

We now assume a thrust density $f_x = 2$ N/cm^2 for the entire stator, and choose the flux-current relationships shown in Fig. 6.11. Referring to (6.14) we choose l_{stroke} = 15 mm initially, so that $(x_o - x_i)/l_{stroke} \approx 0.6$ to 0.8 to have ample time for the current to turn off. Choosing $(x_o - x_i) = 0.8\ l_{stroke}$, and neglecting the coil resistance, from (6.14) we obtain the maximum flux linkage λ_b at the base speed u_b:

$$\lambda_b = \frac{300}{2}\ (0.8 \times 15 \times 10^{-3}) = 1.8 \text{ Wb} \tag{6.25}$$

Next, we choose the primary tooth width/slot opening, $w/b = 1$, and $w = l_{stroke} =$ 15 mm, with the secondary tooth width $b_s = w = 15$ mm (Fig. 6.1). The secondary slot pitch τ_s becomes

$$\tau_s = 2w + b = 2 \times 15 + 15 = 45 \text{ mm} = 3b_s \tag{6.26}$$

Referring to Fig. 6.11, it has been found [6] that the maximum thrust per current increment is produced when $L_s = L_u$, where L_s is the saturated inductance in aligned position and L_u the inductance in unaligned position, as shown in Fig. 6.11. The area in the energy cycle, W, can be found for a given I_b from

$$\frac{W}{k_{safe}} = \lambda_0' \left(\frac{x_o - x_i}{l_{stroke}} \right) + I_b - \frac{\lambda_0'^2}{2L_{au}} \tag{6.27}$$

where L_{au} is the unsaturated inductance in aligned position, $\lambda_0' = \lambda_b - L_s I_b$, (with $L_s = L_u$), and k_{safe} is a safety factor (< 1.0) to account for saturated condition when $L_s \neq L_u$. Next, defining a duty cycle k_d, we write

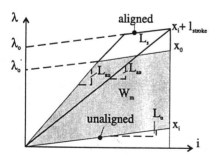

FIGURE 6.11
Simplified λ_i curves for aligned and unaligned positions.

$$\lambda_b = L_{as}I_b \left(\frac{x_o - x_i}{l_{stroke}} \right) = L_{as}I_b k_d \tag{6.28}$$

where L_{as} is the inductance defined in Fig. 6.11 and the subscript b corresponds to base values. Combining (6.27) and (6.28) yields

$$\frac{W}{k_{safe}} = (\lambda_b - L_u I_b)k_d I_b - \frac{(\lambda_b - L_u I_b)^2 k_d^2}{2L_{au}} \tag{6.29}$$

where

$$\lambda_b = L_{as}I_b k_d \tag{6.30}$$

as indicated in (6.28); this can be determined from

$$\lambda_b = 2B_b wlN_1 \tag{6.31}$$

In (6.31) N_1 is the primary number of turns per coil; $l = \pi D_{pi}$, with D_{pi} = primary bore; and B_b is the maximum airgap flux density. The factor 2 appears because we have two teeth per phase in a six-teeth mover (Fig. 6.4), and w is tooth width. In other words, each phase has $2N_1$ turns divided into two coils.

Recall that we have assumed a thrust density $f_x = 2$ N/cm^2 for the entire stator. However, because one-half of the stator surface is occupied by slots, the effective thrust density of the iron surface is $f_{xi} = 2 \times 2 = 4$ N/cm^2. Now, since only two primary teeth are effective in producing the total thrust F_{xb}, we have

$$500 = F_{xb} = \pi D_{pi} \times 2wf_{xi} = \pi D_{pi} \times 2 \times 1.5 \times 4 = 12\pi D_{pi} \tag{6.32}$$

Consequently, $D_{pi} = 13.26$ cm ≈ 15 cm. Next, from (6.25) and (6.31) we obtain

$$N_1 B_b = \frac{\lambda_b}{2wl} = \frac{1.8}{2 \times 0.015 \times \pi \times 0.1326} = 144 \tag{6.33}$$

In order to determine N_1, we express the unaligned inductance L_u as

$$L_u = 2P_{mu}N_1^2 \tag{6.34}$$

where P_{mu} is the permeance in the unaligned position. Similarly the unsaturated inductance L_{au} for the aligned position can be written as

$$L_{au} = 2P_{ma}N_1^2 \tag{6.35}$$

where P_{ma} is the permeance in the aligned position. This permeance is approximately given by

$$P_{ma} \approx \mu_0 \frac{\pi D_{pi} w}{g} \tag{6.36}$$

Finally, the saturated inductance L_{as} in the aligned position is

$$L_{as} = 2P_{mas}N_1^2 \tag{6.37}$$

with

$$P_{mas} = \frac{\mu_0 \pi D_{pi} w}{(1 + k_s)g} \tag{6.38}$$

k_s being the saturation factor. This factor can be determined from a field analysis.

We now assume a maximum airgap flux density $B_b = 1.4$ T for an airgap $g = 0.3$ mm. Thus, from (6.33)

$$N_1 = \frac{144}{1.4} \approx 103 \text{ turns} \tag{6.39}$$

Thus, (6.35), (6.36), and (6.39) yield

$$L_{au} = \frac{2 \times 4\pi \times 10^{-7} \times \pi \times 0.15 \times 0.015 \times 103^2}{0.3 \times 10^{-3}} \approx 0.6 \text{ H} \tag{6.40}$$

From field calculations it has been found [2,4] that

$$L_u = \frac{L_{au}}{10} = \frac{0.6}{10} = 0.06 \text{ H} \tag{6.41}$$

Assuming $k_{safe} = 0.7$, $k_d = 0.8$, and substituting numerical values in (6.29) yields

$$\frac{6}{0.7} = (1.8 - 0.06I_b)0.8I_b - \frac{(1.8 - 0.06I_b)^2}{2 \times 0.6} \times 0.8^2 \tag{6.42}$$

where $W = F_x l_{stroke} k_d = 500 \times 0.015 \times 0.8 = 6$ J. Solving for I_b from (6.42) gives $I_b \approx$ 8 A. The saturated inductance can now be obtained from (6.30) such that

$$L_{as} = \frac{1.8}{8 \times 0.8} = 0.28 \text{ H} \tag{6.43}$$

The required slot area is given by

$$A_{ps} = \frac{N_1 I_b}{J_{co} k_{fill}} \tag{6.44}$$

where J_{co} is the conductor current density, A/m^2, and k_{fill} is a fill factor. Because each phase conducts one-third of the time, the rms base current becomes

$$(I_b)_{rms} = \frac{I_b}{\sqrt{3}} = \frac{8}{\sqrt{3}} = 4.6 \text{ A} \tag{6.45}$$

We choose $(J_{co})_{rms} = 4$ A/mm^2. Thus, $J_{co} = \sqrt{3}(J_{co})_{rms} = \sqrt{3} \times 4 = 6.93$ A/mm^2 and with $k_{fill} = 0.45$ we obtain from (6.44)

$$A_{ps} = \frac{103 \times 8}{6.93 \times 0.45} \approx 264 \text{ mm}^2 \tag{6.46}$$

Since primary slot width $b_p = 15$ mm, slot depth $h_p = 264/15 = 188$ mm.

Turning now to the secondary, the tooth height $h_s = 30g = 30 \times 0.3 = 9$ mm. We recheck the value of the saturation factor k_s from

$$k_s = \frac{L_{au}}{L_{as}} - 1 = \frac{0.6}{0.28} - 1 = 1.14 \tag{6.47}$$

Because the back-core flux lines are the longest, the core thickness may be chosen such that $k_s = 1.14$. Otherwise, the base flux λ_b will be reached at a speed lower than the base speed u_b, and the actuator will not deliver the base thrust at the base speed.

The copper loss is obtained from

$$P_{co} = 3(I_b)_{rms}^2 R_1 = I_b^2 R_1 \tag{6.48}$$

where the primary resistance R_1 is given by

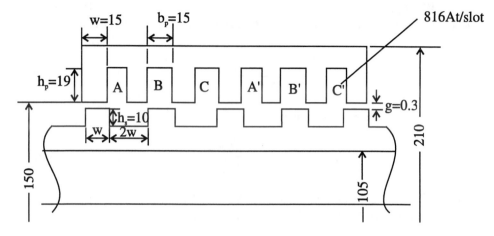

FIGURE 6.12
LSRA geometry for design example.

$$R_1 = 2\rho_{co}\pi(D_{pi} + h_p)N_1^2 \frac{J_{co}}{I_b} \tag{6.49}$$

Substituting $\rho_{co} = 2.3 \times 10^{-8}$ Ωm, $N_1 I_b = 103 \times 8 = 824$ At/slot, $J_{co} = 6.92$ A/mm^2, $N_1 = 103$ turns per coil, $D_{pi} = 0.15$ m, and $h_p = 0.018$ m in (6.49), we obtain $R_1 = 2.15$ Ω. Thus, from (6.48) we get $P_{co} = 8_2 \times 2.15 = 137.6$ W. The core losses can be determined as in rotary switched reluctance motors [1].

Mechanical output power $P_{mech} = F_x u_b = 500 \times 2 = 1000$ W. Finally, the machine geometry is shown in Fig. 6.12.

REFERENCES

1. T. J. E. Miller, *Switched reluctance motors and their control* (Clarendon Press, Oxford, 1993).
2. K. Adamiak et al., "The switched reluctance motor as a linear low-speed drive," *Intl. Conf. on MAGLEV and Linear Drives*, Las Vegas, NV, May 1987, pp. 39-43.
3. P. M. Cusack et al., "Design, control and operation of a linear switched reluctance motor," *Proc. Canada Conf. Elect. and Computer Eng.*, Quebec City, 1991, Paper no. 19.1.1.
4. U. S. Deshpande, et al., "A high force density linear switched reluctance machine," *"IEEE-IAS Rec.*, 1993 Annual Meeting, pt. I, 1993, pp. 251-257.
5. J. Lucidarine et al., "Optimum design of longitudinal field variable reluctance motors," *IEEE Trans.*, vol. EC-8, 1993, pp. 357-361.
6. R. Krishnan et al., "Design procedure for switched-reluctance motors," *IEEE Trans.*, vol. IA-24, 1988, pp. 456-461.

CHAPTER

7

LINEAR STEPPER ACTUATORS

Linear stepper actuators (LSAs) are devices which transform digital (or pulse) inputs into linear incremental motion outputs. They are counterparts of rotary stepper motors. When they are properly operated, the number of linear steps of the actuator equals the number of input pulses, and the mover of the actuator advances by one linear step increment, τ_{ss}. LSAs have applications for short-travel linear step motion. An example of application is the Sawyer motor, which has been successfully applied in drafting equipment, head positioning, laser beam positioning for fabric cutting, etc.

Just like their rotary counterparts, LSAs may be either reluctance or hybrid (with permanent magnets). A unique feature of LSAs is that they can produce directly linear motion insteps smaller than 0.1 mm [1]. Other advantages of LSAs are as follows:

- They can operate as an open-loop system and yet yield a precise position control.
- They are robust and mechanically simple.
- They can be repeatedly stalled without any damage to the actuator.
- Required electronic controllers of LSAs are simple.

  Among the disadvantages of LSAs are as follows;
- They have relatively high losses perthrust (unless the airgap is drastically reduced (to less than 0.2 mm).
- With open-loop operation, the step size is constant.

179

- Step response may have a large overshoot and subsequent oscillations.
- The mover weight tends to be high for the thrust produced.
- Frictional loads and linear ball bearings lead to position errors in open-loop operations.
- Single-sided configurations have large normal forces.
- Special electronics may be required to obtain fast rise and decay of currents as the phase circuits are highly inductive.

Hybrid LSAs, with high-energy product magnets do not possess some of the above-mentioned undesirable features, whereas others may be offset by using closed-loop control.

7.1 PRACTICAL CONFIGURATIONS AND THEIR OPERATION

In the following we will use the acronym VR-LSA for the variable reluctance linear stepper actuator and HPM-LSA for the hybrid permanent magnet LSA. Just like the actuators discussed in the preceding chapters, LSAs may be flat single-sided or double-sided, or they may be tubular and may have three or four phases. An HPM-LSA generally has four phases.

Figure 7.1 shows a double-sided VR-LSA with the three phases a, b, and c on the primary. The primary and the secondary have the same slot pitch τ_s, and the tooth width is equal to the slot opening. For its operation, one or two phases of an LSA may be energized at a given instant. Here, we assume that only one phase is energized at a time. The primary teeth, which constitute the poles, are displaced by τ_s/m ($m = 3$, for three-phase) with respect to the secondary teeth (Fig. 7.1). The step of linear motion is $\tau_{ss} = \tau_s/m$. The primary phases are supplied sequentially, one by one, with unipolar controlled current pulses at a frequency f_s. The mover speed u_s is related to this frequency by

$$u_s = \tau_{ss} f_s \qquad (7.1)$$

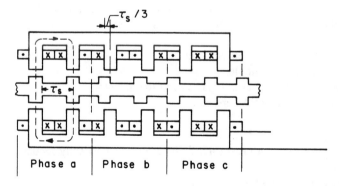

FIGURE 7.1
Double-sided flat VR-LSA.

If the frequency is small enough, the LSA will start from any position. In general, however, for the position of the secondary shown in Fig. 7.1, energizing phase b or c will result in an alignment of the secondary with the primary. Phase a will not produce any thrust, since the mover is in a minimum reluctance position for phase a. The thrust is obtainable from

$$F_x(x) = \frac{\partial W_m'}{\partial x} \tag{7.2}$$

where W_2' is the phase coil coenergy. Or if saturation is neglected (7.2) takes the form

$$F_x(x) = \frac{1}{2} i^2 \frac{\partial L}{\partial x} \tag{7.3}$$

For mechanical reasons, the three phases of the LSA share a common back iron. But the unipolar current polarities in the adjacent phases, being opposite, tend to reduce the magnetic flux interactions between the phases during commutation. As with other actuators having iron in the secondary, normal forces exist in single-sided configurations. In double-sided structures normal forces are zero unless the airgaps on the two sides are unequal. The existence of forces along only one-third of the primary, when only one phase is conducting, leads to noise and vibration. With two phases conducting at a time the situation improves and the step is halved.

An HPM-LSA configuration is shown in Fig. 7.2 [2]. This actuator has four poles and four phases a, b, c, and d. As can be seen from Fig. 7.2, three teeth constitute a pole to reduce the step size to $\tau_s/4$ for a fairly large simple coil per phase. Again, one phase is turned on at a time. For the relative positions shown, phase a is in the position of maximum reluctance and phase b sees the minimum reluctance. In this position, no thrust will be produced if either phase is energized. The minimum reluctance position is called the *detente* position and corresponds to stable equilibrium. The thrust experienced by the secondary, when a phase is energized, has three components. The first is the reluctance force F_{xr}, which is similar to that in a VR-LSA and is produced by virtue of variable reluctances of the primary and the secondary. The second component is the zero current, or detente, force F_{xd} originating from the magnets only. Both these forces are small in comparison with the force F_{xi}, the interaction force, originating from the interaction of the magnet flux λ_{PMj} in a phase and the current in that phase. This force is given by

$$F_{xi} = \left(\frac{\partial \lambda_{PMj}}{\partial x} \right) i_j \tag{7.4}$$

For the position shown in Fig. 7.2, this force is zero, since $\partial \lambda_{PM}/\partial x = 0$ even though i_a and/or i_b may not be zero.

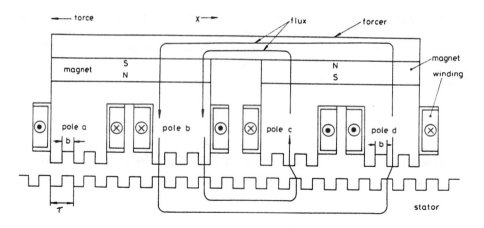

FIGURE 7.2
Single-sided flat HPM-LSA.

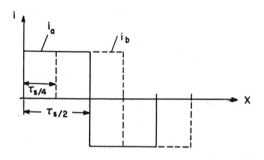

FIGURE 7.3
Ideal bipolar current pulses feeding an HPM-LSA.

For the position shown in Fig. 7.2, the direction of the force depends on the direction of current in phase c or d. So for a bidirectional operation, the static power converter must provide bipolar current pulses. Moreover, because the PM flux path is longer in the end poles, the forces produced by phases a and d are smaller than those of phases b and c (for the same current in each phase). Also, ripples exist in the resultant force, especially at high currents when the core saturates. Finally, from the polarities of currents shown in Fig. 7.2, phases a and b may be connected in series to form a single phase A. Similarly, phases c and d may be connected in series to form a phase B, resulting essentially in a two-phase actuator, which may be supplied with the ideal current pulses shown in Fig. 7.3 These currents are displaced from each other corresponding to the motion step $\tau_{ss} = \tau/4$.

HPM-LSAs may also be made in the form of tubular structures. Reference [3] gives a good description of an HPM-LSA. Tubular HPM-LSAs are particularly suitable for short-stroke applications.

7.2 STATIC FORCES

The finite-element method (FEM) has been used extensively [3,4] to determine the fields and forces pertaining to HPM-LSAs. From these studies it has been found that the outer poles have lower flux densities than those of inner poles, and thereby produce lower forces. Consequently, force pulsations occur, which may result in positioning inaccuracies. Measures have been taken to remedy this. One approach to reducing force pulsations owing to weak end poles is to place the magnet on the secondary [5]. Using a tubular configuration, four modes in the operation of an HPM-LSA with the magnet on the secondary are shown in Fig. 7.4. This device has two ring-shaped coils with independent magnetic circuits on the primary. For all four steps of a cycle, with one phase conducting at a time, the PM fluxes encounter the same permeance. Consequently, the maximum static forces owing to the outer and inner

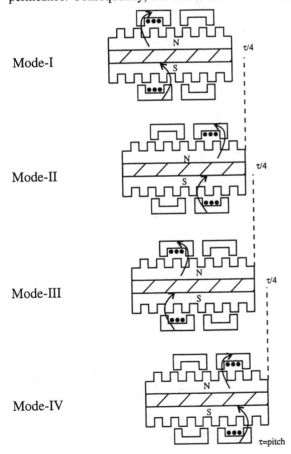

FIGURE 7.4
An HPM-LSA with PM secondary.

poles are almost equal [5]. The principal disadvantage of this configuration is that if the primary is the mover the PM secondary must be all along the travel length. So this type of actuator is suitable for short-stroke applications.

7.3 LUMPED PARAMETER MODELS

Lumped parameter models of LSAs are useful in their dynamic analysis [6]. To illustrate the procedure we consider an HPM-LSA where the phases are magnetically decoupled. The volt-amp equation for a phase is

$$v_k = R_k i_k + \frac{d\lambda_k}{dt} \tag{7.5}$$

In (7.5) the flux linkage λ_k with phase k is

$$\lambda_k(i_k, x, g) = L_k i_k + \lambda_{PMk}(i_k, x, g) \tag{7.6}$$

where the inductance L_k of phase k is given by

$$L_k = L_{km} + L_\sigma \tag{7.7}$$

The inductances L_{km} and L_σ are, respectively, the magnetizing and leakage inductances, and the flux linkages are functions of the phase current i_k, the position x, and the airgap g; also, λ_{PMk} is the flux linkage with phase k due to the magnet flux.

Now, if the actuator has m phases, the thrust F_x and the normal force F_{na} are, respectively, given by

$$F_x \approx \sum_{k=1}^{m} \left(\frac{1}{2} \frac{\partial L_k}{\partial x} i_k + \frac{\partial \lambda_{PMk}}{\partial x} \right) i_k \tag{7.8}$$

and

$$F_{na} \approx \sum_{k=1}^{m} \left(\frac{1}{2} \frac{\partial L_k}{\partial g} i_k + \frac{\partial \lambda_{PMk}}{\partial g} \right) i_k \tag{7.9}$$

Knowing these forces, we can write the mechanical equations as

$$M \frac{d^2 x}{dt^2} = F_x - F_{xload} - C_{fx} \frac{dx}{dt} \tag{7.10}$$

$$M \frac{d^2g}{dt^2} = F_{na} - F_{nload} - C_{fg} \frac{dg}{dt} \tag{7.11}$$

$$\frac{dx}{dt} = u \tag{7.12}$$

$$\frac{dg}{dt} = u_g \tag{7.13}$$

where M is the mass of the mover; C_{fx}, C_{fg} are friction coefficients in x and g (or normal) directions; u, u_g are velocities in the x and g directions; and F_{xload}, F_{gload} are static loads in the respective (x and g) directions. Even if the airgap g is constant, a knowledge of F_{na} is required for designing the mounting frame.

In order to determine the magnetizing inductance L_{km}, we assume that the thrust varies sinusoidally with position for a constant current. From (7.8) it follows that L_k or L_{km} and λ_{PMk} also vary sinusoidally with position. We also assume that L_{km} and λ_{PMk} decrease linearly with current, and note that L_{km} is inversely proportional to g and λ_{PMk} varies inversely with $\sqrt{g}$. Thus, the magnetizing inductances for the four phases can be written as

$$L_{km}(i_k, x, g) = \frac{n_{sp} g_o}{2g} l N_1^2 (1 - C_i i_k) \left[P_0 + P_1 \cos \left(\frac{2\pi x}{\tau_s} - j \frac{\pi}{2} \right) \right] \tag{7.14}$$

where $k = 1, 2, 3, 4$ for the four phases; $j = 0, 1, -2, -1$ for phases 1, 2, 3, and 4, respectively; n_{sp} is the number of slot pitches (τ_s) per pole; l is stack width; and P_0 and P_1 are permeances related to the maximum and minimum permeances P_{max} and P_{min} by

$$P_0 = \frac{1}{2} \left(P_{max} + P_{min} \right) \tag{7.15}$$

and

$$P_1 = \frac{1}{2} \left(P_{max} - P_{min} \right) \tag{7.16}$$

These permeances are shown in Figs. 7.5 and 7.6. Based on conformal mapping, the maximum permeance is given by

$$P_{max} = \frac{\mu_0}{g} \left[w_t + (1 - \sigma) w_s \right] \tag{7.17}$$

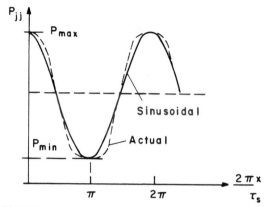

FIGURE 7.5
Airgap permeance variation with position.

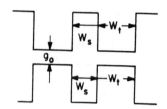

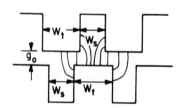

FIGURE 7.6
Geometrics for maximum and minimum permeances.

where

$$\sigma = \frac{2}{\pi}\left[\tan^{-1}\left(\frac{w_s}{g}\right) - g_{2w_s} \ln\left(1 + \frac{w_s^2}{g^2}\right)\right] \qquad (7.18)$$

Similarly, the minimum permeance can be expressed as

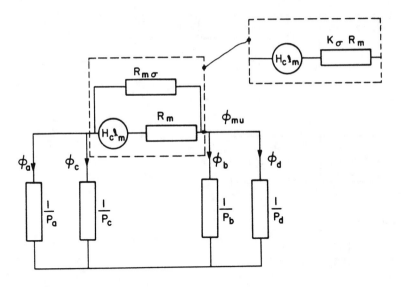

FIGURE 7.7
Magnetic equivalent circuit for PM flux calculation.

$$P_{min} = \frac{4\mu_0}{\alpha} \ln\left(1 + \frac{w_t}{2g}\right)$$ (7.19)

where $\alpha = 1.0$ for $w_s/g < 10$ and $\alpha = 1.1$ for $w_s/g > 10$.

The PM flux $\lambda_{PMk}(i_k, x, g)$ can be determined in terms of k_1, k_2, k_3, and k_4 from

$$\lambda_{PMk}(i_k,x,g) = n_{sp}lN_1(k_1 - k_2)i_k$$

$$\times \left(k_3 + k_4 \cos\left[\frac{2\pi x}{\tau_s} - (k - 1)\frac{\pi}{2}\right]\right)\sqrt{\frac{g_0}{g}}$$ (7.20)

and the magnetic equivalent circuit shown in Fig. 7.7, where

$$R_m = \frac{l_m}{\mu_{re}h_m}$$ (7.21)

is the reluctance of the PM per unit stack width. Other symbols in (7.21) are l_m, PM length; h_m, PM height; and μ_{re}, PM recoil permeability.

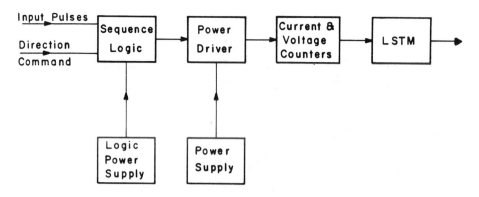

FIGURE 7.8
Basic open-loop control system.

7.4 CONTROL ASPECTS

A drive control, in general, consists of a control power amplifier that sequentially energizes the phase windings of the motor [7]. The power amplifier may be switched unidirectionally or bidirectionally by logic circuitry. Basic elements of the motor control, as shown in Figure 7.8, are the sequence logic, power driver, power supply, and current and voltage limiters. The sequence logic receives input step and direction signals and converts them into base-drive signals for the power transistor in the power driver. The power transistors amplify the power to energize the motor. To limit the voltage spike on power transistors to a safe value, when a phase is de-energized, a suppression circuitry is provided. The control system of Fig. 7.8 is a typical open-loop system. However, position, speed, or current feedback may be added to the system.

Logic Sequencers

The logic sequencers for LSAs, as presented here, are built for a one- or two-phase-on operation. For a four-phase motor, the timing diagrams of the one-phase-on and two-phase-on operation are respectively shown in Figs. 7.9(a) and (b). Similar timing diagrams are valid for three-phase LSAs.

Numerous types of logic circuits are commercially available. A typical four-phase one-phase-on logic sequence, using binary counting technique and two flip-flops, for unidirectional motion is shown in Fig. 7.10. The circuit consists of a conventional up-counter made of two J-K flip-flops. The AND gates decode the counter to provide the four drive signals for the output. As only one phase is on at a time, the sequence of switching is *abcda* . . . . The counter will start up in the RESET condition when power is applied due to the RC network connected to the direct reset inputs of the flip-flops. By adding gating networks that control the direction of counting, the unidirectional motion sequences can be converted to bidirectional mode.

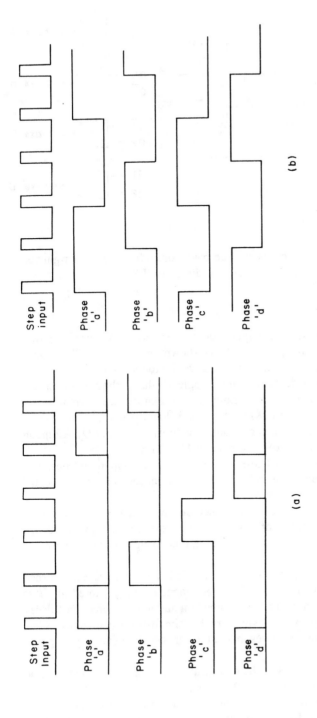

FIGURE 7.9
Timing diagrams: (a) one phase on; (b) two phases on.

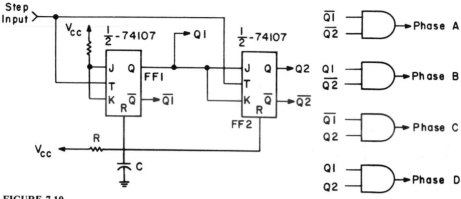

FIGURE 7.10
Logic sequencer for one phase on.

The timing diagram is still valid, but should be read from left to right for rightwise direction (RW) and from right to left for leftwise direction (LW).

Power Drivers

The power driver may be built as a unipolar or bipolar driver. A unipolar power driver, for one phase of a three-phase LSA, is shown in Fig. 7.11. The power amplifier, driven by TTL logic, utilizes a standard Darlington switch.

When the logic sequencer delivers its high (logic 1), the Darlington transistors Q_1 and Q_2 are turned on through R_1. As Q_2 saturates, the motor phase current passes from the source through Q_2 to ground. When the logic is low (logic 0), the base drive to Q_1 is shunted to ground through the logic gate. This turns off Q_1 and Q_2. A bipolar drive circuit for a four-phase HPM-LSA is given in Fig. 7.12.

Two of the major problems of LSAs are related to the buildup and decay of phase currents as phases are turned on and off. When the frequency of the input signal increases, the phase current does not have enough time to build up, or decay, as the motor phase circuit is highly inductive. When turned off, the voltage across the power transistor is $L_m(di/dt)$. The magnitude of this voltage may exceed the V_{ce} rating of the transistor and may cause it to break down. Therefore, this voltage must be suppressed by a suppression circuit [1].

In an active suppression circuit, it should be noticed that when phase a is turned off, phase b is turned on simultaneously. Thus, the magnetic energy stored in phase a is transferred to the phase that is just being turned on and to the source. Fast decay rates of currents and significant energy savings are thus simultaneously obtained. Also, the open-loop range of operation increases, but at the expense of higher step-response oscillations.

As the switching speed increases, there is insufficient time for the current and the thrust to build up to sustain the motion. Overdriving methods are used to increase the speed range. Among the most common methods are (a) series-resistance driver; (b) dual-voltage control; (c) chopped-voltage control; (d) active-suppression control (just discussed); and (e) voltage-multiplying drive.

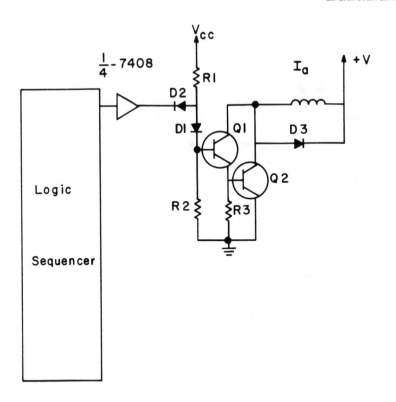

FIGURE 7.11
Unipolar current power drive.

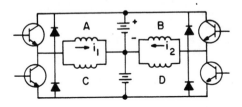

FIGURE 7.12
Bipolar drive circuit for four phases.

LSA Control

An LSA operates in either the open-loop or the closed-loop mode. These are now briefly discussed.

Open-Loop Operation

The open-loop operation may be further classified into *stepping mode* and slewing *mode*. In the stepping mode, the motor stops at every step, before the next phase is

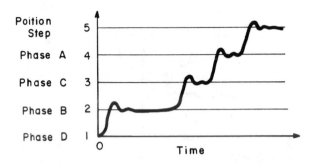

FIGURE 7.13
Step response for open-loop control.

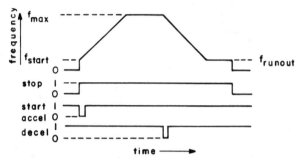

FIGURE 7.14
Linear ramp technique for positioning.

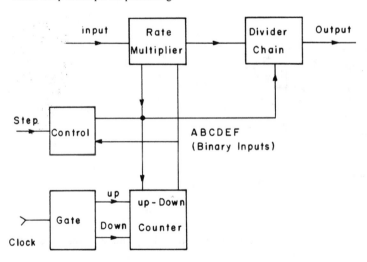

FIGURE 7.15
All-digital linear ramp controller.

switched on to initiate the next step. A typical response in the stepping mode is shown in Fig. 7.13.

For open-loop slewing mode operation, an acceleration-deceleration gated pulse program is used to avoid losing steps. This means that, in the open-loop mode, the actuator is started at a pulse rate within its stepping rate. Then the pulse rate is gradually increased until slewing speed is reached, and the motor is decelerated to a speed within its start-stop range, from which it can be stopped on command without position error.

There are many possible open-loop acceleration-deceleration schemes, such as linear ramp, RC ramp, exponential ramp, and digital techniques. The linear ramp technique is illustrated in Fig. 7.14. The open-loop acceleration-deceleration profile can be obtained by analog or digital techniques. An all-digital system having rate amplifiers used for linear acceleration-deceleration profile with rate multipliers controlled by up-down counters is shown in Fig. 7.15. As the actuator is not supposed to lose steps, it follows the profile of Fig. 7.13. An overview of open-loop control of stepper motors, also valid for LSA, is given in Reference [8].

Closed-Loop Control

The LSA closed-loop control is very similar to that of a rotary stepper motor. The angular displacement is replaced by a linear displacement and the torque by thrust. The extensive literature may be consulted on the subject [9]. Further investigations should include damping of LSAs, magnetic linear bearings, linear position encoders for closed-loop control, new topologies, and digital-control strategies. The application of a linear bearing also implies motion along g_0. Damping of LSAs should include new friction, viscous, eddy current, and electronic damping [7].

7.5 LSA DESIGN GUIDELINES

The design of an LSA drive system includes the following:

- design of LSA;
- power driver selection;
- control system selection;
- performance simulation; and
- final requirements for optimum performance.

The last four items in this list are similar to those for rotary stepper motors and are discussed in detail in Reference [7]. Here we consider only the design of a VR-LSA. The following data are pertinent to the actuator:

linear step size τ_{ss} = 1 mm
total length of stroke l_s = 100 mm
load weight M_g = 1.5 kg
load friction coefficient $c \sim 0$
capacity to accelerate in either direction from 1 to 50 increments (along 50 steps) and come to a complete stop (fully damped) within 300 ms

As the first step in design, we calculate the peak static thrust F_{axp}. For a constant acceleration a, with $c = 0$, we have

$$F_{axp} = (M_g + M_{actuator})a \qquad (7.22)$$

The acceleration over 50 steps, corresponding to $l_s = 50$ mm, must be complete within $t_a = 100$ ms (out of 300 ms). Hence,

$$a = \frac{2l_s}{t_a^2} = \frac{2 \times 50 \times 10^{-3}}{(100 \times 10^{-3})^2} = 10 \text{ m/s}^2 \qquad (7.23)$$

The maximum attainable speed is

$$u_{max} = at_a = 10 \times 0.1 = 1 \text{ m/s} \qquad (7.24)$$

and

$$f_s = \frac{u_{max}}{\tau_{ss}} = \frac{1}{1 \times 10^{-3}} = 1000 \text{ Hz} \qquad (7.25)$$

Because the actuator weight is not known, let us assume a 2 to 1 thrust-to-weight ratio k_m for the actuator. Therefore,

$$M_{actuator} = \frac{F_{axp}}{k_m g} \qquad (7.26)$$

From (7.23) and (7.22) we get

$$F_{axp} = M_g \left/ \left(\frac{1}{a} - \frac{1}{k_m g} \right) \right. = 1.5 \left/ \left(\frac{1}{10} - \frac{1}{2 \times 9.81} \right) \right. \sim 30.5 \text{ N} \qquad (7.27)$$

The airgap is mechanically limited between 0.1 and 0.2 mm. Let us choose $g_0 = 0.2$ mm, and let the actuator have the structure, for one phase, as shown in Fig. 7.16.

As the step size $\tau_{ss} = 1$ mm and the actuator has three phases, tooth pitch $\tau_s = 3\tau_{ss} = 3$ mm. If tooth width $w_t =$ slot width w_s, we obtain $w_s = w_t = 1.5$ mm.

We now determine the maximum and the minimum values of permeances from (7.17) through (7.19). Hence,

$$\sigma = \frac{2}{\pi} \left[\tan^{-1} \frac{w_s}{g_0} - \frac{g_0}{2w_s} \ln \left(1 + \frac{w_s^2}{g_0^2} \right) \right]$$

$$= \frac{2}{\pi} \left[\tan^{-1} \left(\frac{1.5}{0.2} \right) - \frac{0.2}{2 \times 1.5} \ln(1 + 7.5^2) \right]$$

$$= 0.744 \tag{7.28}$$

$$P_{max} = \frac{\mu_o}{g_0} \left[w_t + (1 - \sigma)w_s \right]$$

$$= \frac{\mu_0}{0.2} [1.5 + (1 - 0.744)1.5]$$

$$= 9.42 \mu_o \tag{7.29}$$

$$P_{min} = \mu_0 \left[\frac{4}{\alpha} \ln \left(1 + \frac{w_t}{2g_0} \right) \right]$$

$$= \mu_0 \left[\frac{4}{1} \ln \left(1 + \frac{1.5}{2 \times 0.2} \right) \right]$$

$$= 6.23 \mu_0 \tag{7.30}$$

From (7.8) and (7.14) through (7.16), the peak thrust F_{axp} becomes

$$F_{axp} = \frac{1}{2} l(NI_{dc})^2 \frac{2\pi}{\tau_s} n_{sp} \left[\frac{1}{4} (P_{max} - P_{min}) \right] \tag{7.31}$$

As the number of teeth per pole $n_{sp} = 3$, the remaining unknowns in (7.31) are l and (NI_{dc}), where l is the active width of the primary stack and NI_{dc} is the total primary mmf per phase. From Fig. 7.15

$$NI_{dc} = 4n_c I_{dc} \tag{7.32}$$

where n_c is the number of turns per coil. There are four coils per phase. The primary mmf per phase can be obtained by setting a limit on the maximum airgap flux density B_{gp} for the tooth-against-tooth position (Fig. 7.16). Hence,

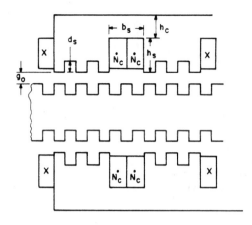

FIGURE 7.16
Longitudinal cross section of a flat double-sided VR-LSA.

$$NI_{dc} \sim \frac{B_{gp}}{\mu_0}(4g_0) \tag{7.33}$$

Thus, for a given value of B_{gp}, n_c can be obtained from (7.32) and (7.33). The area required by the primary coils ($b_s h_s$ in Fig. 7.16) is given by

$$b_s h_s = \frac{NI_{dc}}{2k_{fil}J_{co}} \tag{7.34}$$

where k_{fil}, the fill factor, is 0.5 to 0.6 and J_{co}, the design current density, is 2 A/mm^2 for continuous duty and 4 A/mm^2 for short duty and natural cooling. To reduce the leakage of the primary coils, we take $h_s = 1.5b_s$. With $B_{gp} = 0.5$ T, (7.33) yields

$$NI_{dc} = \frac{0.5 \times 4 \times 0.2 \times 10^{-3}}{4\pi \times 10^{-7}} = 318 \text{ At}$$

From (7.34)

$$h_s b_s = \frac{318}{2 \times 0.5 \times 4} = 79.5 \text{ mm}^2$$

As $h_s = 1.5 \, b_s$, we finally obtain

$$b_s = 7.28 \text{ mm}$$

and

$$h_s = 10.92 \text{ mm}$$

Hence, from (7.32) the peak thrust becomes

$$F_{axp} = \frac{1}{2} l(318)^2 \frac{2\pi}{3 \times 10^{-3}} \times 3$$

$$\times \frac{1}{4} \mu_0 (9.42 - 6.23) = 318.05 l$$

But $F_{axp} = 30.5$ N, as obtained from (7.31). Hence,

$$l = \frac{30.5}{318.05} = 95.8 \text{ mm}$$

The slot depth has not been considered in the permeance calculations. In general, the ratio h_s/g_o should be greater than 20. The core depth h_c (Fig. 7.15) is determined from mechanical constraints, as magnetically

$$h_c \sim 3w_t \frac{B_{gp}}{B_c} = \frac{3 \times 1.5 \times 10^{-3} \times 0.5}{1.0} = 2.25 \text{ mm} \qquad (7.35)$$

where B_c is the core flux density. Considering a high normal (attraction) force between the two primaries of a double-sided actuator, a value of 4 mm is assigned to h_c.

The weight of the actuator is calculated as follows:

Total weight of copper (for the three phases),

$$G_{cot} = 3 \times 2\gamma_{co} h_s b_s k_{fill} (2w_A + 6\tau_s + 2b_s)$$

$$= 3 \times 2 \times 8.9 \times 10^3 \times 79.5 \times 10^{-6}$$

$$\times 0.5(2 \times 95.8 + 6 \times 3 + 2 \times 7.28) \times 10^{-3}$$

$$= 0.475 \text{ kg} \qquad (7.36)$$

Total weight of iron (for the two stators),

$$G_{it} = 3 \times 2\gamma_i w_A [(6\tau_s + 2b_s)h_c + 6w_t d_s + 6\tau_s(h_s - d_s)]$$

$$= 2 \times 2 \times 7.6 \times 10^3 \times 95.8 \times 10^{-3} [(6 \times 3 + 2 \times 7.28)4$$

$$+ 6 \times 1.5 \times 4 + 6 \times 3(10.92 - 4)]10^{-6}$$

$$= 0.847 \text{ kg}$$

Hence, the total weight of the actuator $= 0.475 + 0.847 = 1.322$ kg

The thrust/weight ratio $= 30.5/(1.322 \times 9.81) = 2.352$, which is more than 2, which was the target. To reduce the actuator weight further, let us increase the airgap flux density to 0.7 T without modifying the core height. This increase will require an increase in NI_{dc}, but not so much of the copper weight because the actuator effective width will be drastically reduced. Following the same approach as above, from (7.33), for $B_{gp} = 0.7$ T, $NI_{dc} = 445$ At/phase,

$$h_s b_s = \frac{445}{2 \times 0.5 \times 4} = 111.25 \text{ mm}^2$$

Let $h_s = b_s = 10.54$ mm. From the force equation,

$$F_{axp} = 523.5l = 30.5 \text{ N}$$

Hence, $l = 58.3$ mm. Consequently,

$$G_{cot} = 3 \times 2 \times 8.9 \times 10^3 \times 111.25 \times 10^{-6}(2 \times 58.3$$

$$+ 6 \times 3.3 + 2 \times 7.28)10^{-3} = 0.786 \text{ kg}$$

$$G_{it} = \frac{0.8847}{95.8} \times 49 = 0.433 \text{ kg}$$

Total weight $= 0.786 + 0.433 = 1.219$ kg and

$$k_m = \frac{30.5}{1.219 \times 9.81} = 2.55$$

Now that the main dimensions of the actuator are fixed, the analysis can be refined by advanced field modeling to account for saturation and leakage fields.

REFERENCES

1. S. A. Nasar and I. Boldea, *Linear electric motors* (Prentice-Hall), Englewood Cliffs, NJ, 1987), Chapter 6.
2. T. Yokozuka and E. Baba, "Force-displacement characteristics of linear stepper motors," *Proc. IEE*, vol. B-139, 1992, pp. 37-43.
3. Y. Takeda et al., "Linear pulse motors for accurate positioning," *Rec. IEEE-IAS*, 1991 Annual Meeting, pt. I, pp. 144-149.
4. 1992 Annual Meeting
5. Y. Takeda et al., "Optimum tooth design for linear pulse motors," *Rec. IEEE-IAS*, 1989 Annual Meeting, pt. I, pp. 272-276.
6. I. Viorel et al., "Sawyer type linear motor dynamic modeling," *Rec. ICEM*, 1992, vol. 2, pp. 697-701.
7. B. C. Kuo (editor), *Step motors and control systems* (SRL Publishing Co., Champaign, IL, 1979).
8. A. C. Leenhouts, "An overview of modern step motor control methods," *PCIM Journal*, no. 2, 1993, pp. 54-61.

CHAPTER

8

LINEAR ELECTRIC GENERATORS

In the preceding chapters we have studied linear electric actuators, which convert electrical energy into mechanical form. The operation of such actuators is reversible in that they may be used as linear electric generators to convert mechanical energy into electrical energy. Whereas this statement is correct in principle, because of practical constraints imposed by the availability of the prime mover (such as the free-piston Stirling engine) not every configuration of actuators discussed earlier is suitable for the generator mode of operation. For instance, the stroke of the generator is limited to about 3 cm; three-phase voltage generation (with *abc* phase rotation) is not obtainable owing to the phase reversal in the return stroke; and only ac generation is considered practical. Therefore, in the following we will discuss only single-phase linear ac generators or linear alternators (LAs) having short strokes of motion [1,2].

8.1 PRINCIPLE OF OPERATION AND BASIC CONFIGURATIONS

Just like their rotary counterparts, LAs operate on the basis of Faraday's law of electromagnetic induction, according to which

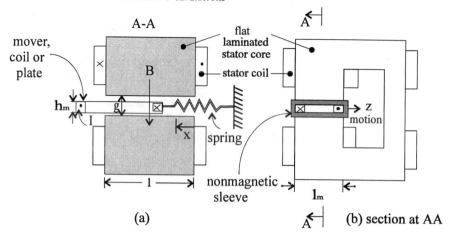

FIGURE 8.1
A moving coil linear alternator.

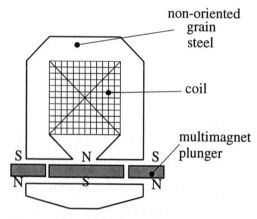

FIGURE 8.2
A moving magnet linear alternator.

$$e = \frac{d\lambda}{dt} \tag{8.1}$$

That is, a voltage e is induced in a coil if its flux linkage λ varies with time t. Consequently, an LA may belong to one of the following categories:

(i) Moving coil type
(ii) Moving magnet (or electromagnet) type
(iii) Moving iron type

These various types have advantages and disadvantages, and are schematically represented in Figs. 8.1, 8.2, and 8.3 respectively.

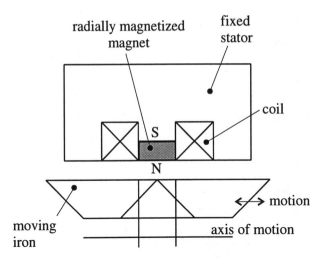

FIGURE 8.3
A moving iron linear alternator.

The moving coil LA requires flexible leads, which tend to wear out, especially in high-power machines. An example of the moving coil permanent magnet (PM) LA is the PM moving coil loudspeaker. However, for any significant power generation the moving coil LA is not suitable. Notice from Fig. 8.1 that such an LA inherently has a large airgap.

Some of the shortcomings of a moving magnet LA are the following: large magnet leakage fields; exposure of PMs to continuous vibrations, which may demagnetize the magnets; constraints on power conversion capabilities by the stroke length and the length of the magnet; thickness of the magnets as dictated by the airgap; and size of the magnets as limited by the mass of the moving member of the LA.

Moving iron LAs are rugged and have other advantages, but they tend to be relatively heavier. In the following we discuss moving magnet and moving iron LAs. The moving coil LA, being of little practical interest, will not be considered further.

8.2 MOVING MAGNET LINEAR ALTERNATORS

Whereas numerous configurations of moving magnet LAs are feasible, two topologies are shown in Figs. 8.2 and 8.4. The first (Fig. 8.2) has multiple magnets, whereas the second (Fig. 8.4) has a single moving magnet. Clearly, there is no difference in their principles of operation. For our discussion and analysis we will consider the LA with a single moving magnet of Fig. 8.4. (*Note*: The end magnets of small thickness are used as "spring" magnets to restore the mover to the center position.)

The first step in the analysis and design of a permanent magnet linear alternator (PMLA) is the determination of the airgap field produced by the magnet. Whereas the

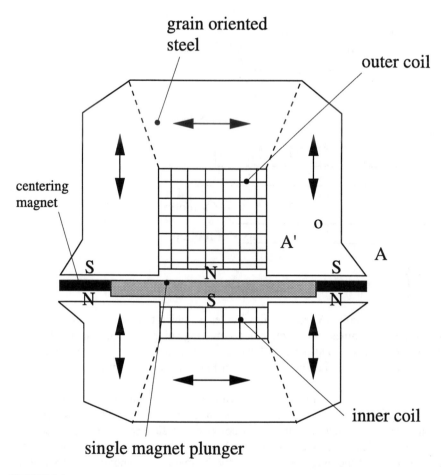

FIGURE 8.4
A single magnet linear alternator with spring magnets.

fields may be precisely determined by finite-element analysis, a magnetic circuit approach yields approximate results acceptable for practical purposes. Thus, neglecting saturation and leakage, for the magnet position shown in Fig. 8.5, we have

$$B_m = B_g = \mu_0 H_g \tag{8.2}$$

Neglecting the reluctance of the core, Ampere's law yields

$$k_s H_g (2g) + H_m h_m + H_m h_s = 0 \tag{8.3}$$

Assuming rare-earth magnets, their demagnetization characteristic can be written as

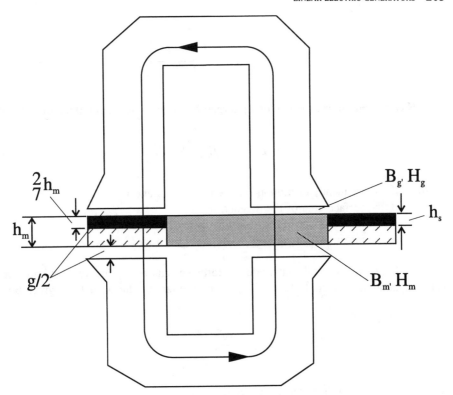

FIGURE 8.5
Magnet in one extreme position.

$$B_m = B_r + \mu_{rc}H_m \tag{8.4}$$

The various symbols in (8.3) and (8.4) are defined as follows:

B_m = PM operation flux density, T
H_m = PM operation field intensity, A/m
g = airgap, m
k_s = saturation factor
B_r = residual flux density, T
μ_{rc} = PM recoil permeability, H/m
B_g = airgap flux density, T
H_g = airgap magnet field intensity, A/m
h_m = main magnet thickness, m
h_s = spring magnet thickness (being backed by iron core)

Combining (8.2) through (8.4) yields

$$B_m = B_r \left[1 + \frac{\mu_{rc}}{\mu_0} \left(\frac{2k_s g}{h_m + h_s} \right) \right]^{-1}$$ (8.5)

Having determined the maximum airgap flux density, we write the induced emf
E

$$E = 4.44 f B_m A_g N$$ (8.6)

where f = frequency of motion (of the moving magnet)
A_g = surface area of the magnet, m²
N = total number of turns on the stator
B_m = maximum airgap flux density, T [given by (8.5)]

In order to know the alternator performance characteristics we must know the various parameters of the PMLA. First, the magnetizing inductance L_m is given by

$$L_m = \frac{\mu_0 \pi D_m N^2 l_s}{2k_s g_m}$$ (8.7)

where D_m = mean diameter of the windings, m
l_s = stroke length, m
g_m = magnetic airgap, m

and other symbols have been defined earlier. Next, referring to the symbols shown in Fig. 8.6, the leakage inductance is given by

$$L_\sigma = \mu_0 N_i^2 \left(\frac{h_{si}}{3b_{si}} + \frac{h_s}{b_{si}} \right) \pi D_i + \mu_0 N_0^2 \frac{h_{so}}{3b_{so}} \pi D_0$$ (8.8)

Finally, the synchronous inductance L_s becomes

$$L_s = L_m + L_\sigma$$ (8.9)

Since $N = N_i + N_0$ and $A_g = \pi D_m l_s$, we may rewrite (8.6) as

$$E = 4.44 \pi f (N_i + N_0) B_m D_m l_s$$ (8.10)

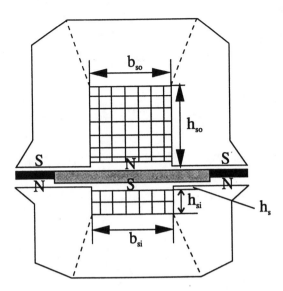

FIGURE 8.6
Alternator geometry and dimensions.

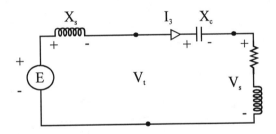

FIGURE 8.7
Equivalent circuit of a PMLA.

In order to obtain a relatively good voltage regulation, a "tuning" capacitor is sometimes used to offset the synchronous reactance. The equivalent circuit of the PMLA then becomes as shown in Fig. 8.7. With a tuning capacitor in the circuit we proceed to determine the operating condition of a given alternator.

Example 8.1 A 80-kW 60-Hz PMLA has the following data: inner winding turns $N_i = 16$; outer winding turns $N_0 \approx 36$; mean diameter of inner winding $D_i = 0.33$ m; mean diameter of outer winding $D_0 = 0.3986$ m; magnet residual flux density $B_r = 1.1$ T; magnet recoil permeability $\mu_{rc} = 1.05 \mu_0$; main magnet thickness $h_m = 7$ mm; spring magnet thickness $h_s = 2$ mm; mechanical airgap $g = 1$ mm; stroke length $l_s = 0.04$ m. Other dimensions shown in Fig. 8.6 are as follows: $h_{si} = 10$ mm; $h_s = 6$ mm; $b_{si} = 45$ mm; $h_{so} = 32$ mm. Tuning capacitor $C = 512$ μF. If the alternator supplies full-load current (147-A) at 0.85 lagging

lagging power factor, determine the terminal voltage at this load. Neglect the winding resistance, and assume saturation factor $k_s = 1.1$.
From (8.5) we have

$$B_m = 1.1 \left(\frac{1 + 1.05 \times 2 \times 1.1 \times 1}{7 + 2} \right)^{-1} = 0.8753 \text{ T}$$

From (8.10) the induced voltage becomes

$$E = 4.44\pi \times 60(16 + 36)0.8753 \times 0.364 \times 0.04 = 554 \text{ V}$$

Next, we determine the alternator synchronous reactance. From (8.7) and the given data we have

$$L_m = \frac{4\pi \times 10^{-7} \times \pi \times 0.364 \times (36 + 16)^2 \times 0.04}{2 \times 1.1 \times 8 \times 10^{-3}} = 8.83 \text{ mH}$$

which is the magnetizing inductance. The leakage inductance is obtained from (8.8) as

$$L_\sigma = 4\pi \times 10^{-7} \times 16^2 \left(\frac{10}{3 \times 45} + \frac{6}{45} \right) \pi \times 0.33$$

$$+ 4\pi \times 10^{-7} \times 36^2 \times \frac{32}{3 \times 46} \times \pi \times 0.3986 = 0.542 \text{ mH}$$

Thus, (8.9) yields

$$L_s = 8.83 + 0.542 = 9.372 \text{ mH}$$

At 60 Hz the synchronous reactance is

$$X_s = 2\pi \times 60 \times 9.372 \times 10^{-3} = 3.53 \ \Omega$$

The tuning capacitor of 512 µF capacitance has a capacitive reactance of

$$X_c = \frac{10^6}{2\pi \times 60 \times 512} = 5.18 \ \Omega$$

With $X_c > X_s$ we draw the phasor diagram and the equivalent circuit, as shown in Fig. 8.8, from which we obtain

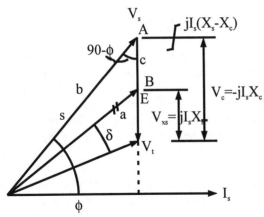

FIGURE 8.8
Phasor diagram of a PMLA.

$$V_s^2 + I_s^2(X_c - X_s)^2 - 2V_s I_s (X_c - X_s) \cos(90 - \phi) = E^2$$

Substituting various numerical values in this equation we obtain V_s = 642 V, which is the terminal voltage across the load. The output power is

$$P_0 = V_s I_s \cos \phi_s = 642 \times 147 \times 0.85 \approx 80 \text{ kW}$$

In this example we have neglected the coil resistance. However, for efficiency calculations, this resistance must be known. This resistance is given by

$$R_s = \rho_{co}\left(\frac{l_i N_i^2}{A_{coi}} + \frac{l_o N_o^2}{A_{coo}}\right) \qquad (8.11)$$

where $A_{coi} = A_i k_{fill}$, $A_{coo} = A_o k_{fill}$, l is the mean length per turn, ρ_{co} us the resistivity of copper, A is the area of conductor cross section, and subscripts i and o, respectively, correspond to inner and outer coils.

For the alternator, we let A_{coi} = 506 mm² and A_{coo} = 1518 mm². Choosing a fill factor of 0.8, substituting the following numerical values in (8.11)

ρ_{co} = resistivity of copper = 2.13 × 10⁻⁸ Ωm at 80°C
l_i = πD_i = $\pi(0.33)$ = 1.0456 m
l_o = πD_o = $\pi(.0398)$ = 1.25 m
N_i = 16 and N_o = 36
yields
R_s = 0.0424 Ω

8.3 MOVING IRON LINEAR ALTERNATORS

Just as with moving magnet LAs, numerous configurations of LAs are feasible where magnets are stationary and "iron" constitutes the moving member. For our discussion and to illustrate the operating principle, analysis, and design, we consider the topology shown in Fig. 8.9. Here we show that the mover is made of two parts, although the mover may have only one segment or more than two segments. The two mover segments shown in Fig. 8.9 are separated from each other by a distance l_s by using a lightweight nonmagnetic spacer. The stator is stacked together as four units. The PMs are placed in the airgap as shown.

The magnet flux in each of the four stator segments varies from a minimum flux ϕ_{min} to a maximum flux ϕ_{max}. For two stator segments, the magnet flux is positive (say, of N-polarity) and negative in the other two, as shown in Fig. 8.10. The total flux per pole linking the coil varies between $+2(\phi_{max} - \phi_{min})$ and $-2(\phi_{max} - \phi_{min})$. In the middle, the total flux linking the coil is zero. With negligible saturation, we may assume a linear variation of flux with the mover position (Fig. 8.10), and assuming a harmonic motion we have

$$x = \frac{1}{2} l_s \sin \omega_1 t \tag{8.12}$$

where l_s is the stroke length and ω_1 the frequency of oscillation. The total induced voltage, with the four coils connected in series, is given by

$$e_a = 8(\phi_{max} - \phi_{min}) N_c \omega_1 \cos \omega_1 t \tag{8.13}$$

where N_c is the number of turns per coil. The rms value of the induced voltage is

$$E_a = 35.538 f_1 (\phi_{max} - \phi_{min}) N_c \tag{8.14}$$

For a maximum voltage, ϕ_{min} must be reduced as much as possible. For the configuration under consideration, this goal is accomplished when

$$g + h_m \approx \frac{1}{2} l_s \tag{8.15}$$

where g is the mechanical airgap and h_m the magnet thickness. Now, to determine ϕ_{max} and ϕ_{min}, we refer to the permeance functions G's shown in Fig. 8.11 for two mover positions. The permeance G_4 is zero; and, for all practical purposes, only G_3 remains for the first mover segment and $2G_3$ for the second. However, G_3 on the left side of mover segment 1 cancels one G_3 for the ϕ_{min}. Finally, G_{2m} cancels G_5. Thus, for the difference ($\phi_{max} - \phi_{min}$) only $G_{g1}(0)$ counts for ϕ_{max} and G_3 for ϕ_{min}.

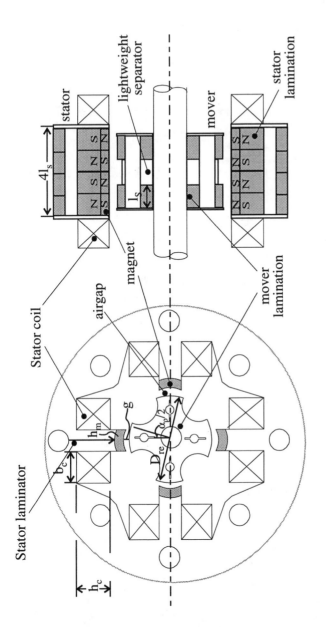

FIGURE 8.9
A novel moving iron linear alternator.

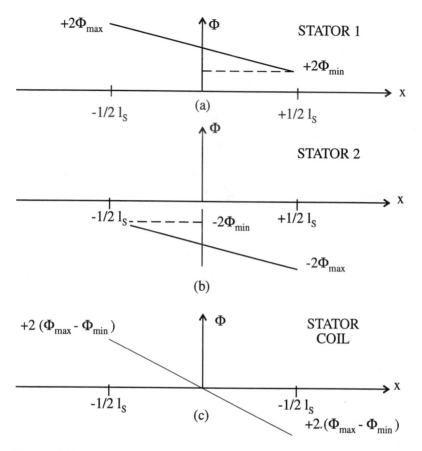

FIGURE 8.10
Permanent magnet stator fluxes (per pole) versus mover position: (a) stator no. 1; (b) stator no. 2; and (c) stator coil.

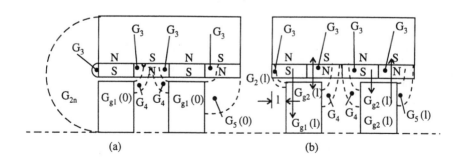

FIGURE 8.11
Permeances for (a) $l = 0$ and (b) $l \neq 0$.

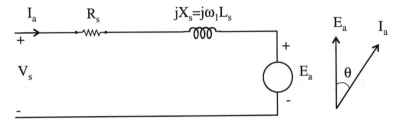

FIGURE 8.12
Equivalent circuit for actuator operation.

To proceed with the flux determination, we replace the PM by its equivalent mmf I_{PM} such that

$$I_{PM} = \frac{B_r h_m}{\mu_{rc}} \qquad (8.16)$$

where $\mu_{rc} \approx 1.05\,\mu_0$ = magnet recoil permeability and B_r the remnant flux density. This mmf is at the airgap and sees the permeances $G_{g1}(0)$ and G_3. Therefore, the maximum and minimum fluxes are given by

$$\phi_{max} = G_{g1}(0)I_{PM} \qquad (8.17)$$

$$\phi_{min} = G_3 I_{PM} \qquad (8.18)$$

where the permeances are expressed as

$$G_{g1}(0) = \mu_0 \pi \left[D_{re} + \frac{1}{2}(g + h_m) \right] \left(\frac{l_s}{g + h_m} \right) \frac{\alpha_p}{360} \qquad (8.19)$$

$$G_3 = 3.3\mu_0 \left[\frac{1}{2}D_{re} + \frac{1}{2}(g + h_m) \right] \frac{\alpha_p}{360} \qquad (8.20)$$

The symbols in these expressions are defined in Fig. 8.9.
As was done for the moving magnet LA, we now develop the equivalent circuit (Fig. 8.12). The parameters of this circuit are

$$R_s = \frac{4N_c l_c \rho_c}{A_c} \qquad (8.21)$$

which can also be written as

$$R_s = \frac{4\rho_c l_c N_c^2 J_c}{N_c I_a} \tag{8.22}$$

where ρ_c is the conductivity of copper, l_c the mean length per turn, A_c the conductor cross section, N_c the number of turns per coil, J_c the design current density, and I_a the rated current.

We define the coil permeance G_{coil} by

$$G_{coil} = 2[G_{g1}(0) + G_3] \tag{8.23}$$

The magnetizing inductance is given by

$$L_m = 4 G_{coil} N_c^2 \tag{8.24}$$

Finally, the leakage inductance has two components. These are slot leakage and end-connection leakage inductances. Thus, the total leakage inductance becomes

$$L_\sigma = 8N_c^2(\lambda_s l_{cs} + \lambda_e l_{ce}) \tag{8.25}$$

where $l_{cs} = 4l_s$ = coil length in the slot; l_{ce} = coil end-connection length per side = $1/2(l_c - 2l_{cs})$; $\lambda_e = 1/2\lambda_s$ = end-connection-specific permeance; and λ_s = slot-specific permeance, which is given by

$$\lambda_s = \mu_0 \left\{ h_1 \Big/ \left[\frac{1}{2}(b_c + b_{c1}) \right] + h_2 \Big/ \left[\frac{1}{2}(b_{c1} + b_{c2}) \right] \right\} \tag{8.26}$$

where the various dimensions are shown in Fig. 8.9. Combining (8.24) and (8.25) yields the total stator inductance:

$$L_s = L_m + L_\sigma \tag{8.27}$$

Consequently, we obtain the parameters of the LA equivalent circuit shown in Fig. 8.12.

We now present the procedure for designing a moving iron LA by an illustrative example.

Example 8.2 The goal is to design an LA of the type shown in Fig. 8.9 having the following specifications: output power P_{oe} = 1000 W (supplying a resistive load with a tuning capacitor in the circuit); rated output voltage V_n = 120 V;

rated frequency f_n = 60 Hz; stroke length l_s = 20 mm; desired efficiency η_n = 0.92. We wish to determine the physical dimensions and other design data.

Rated current, $\quad I_n = \dfrac{P_{oe}}{V_n} = \dfrac{1000}{120} = 8.33$ A

Developed electromagnetic power, $\quad P_e = \dfrac{P_{oe}}{\eta_n} = \dfrac{1000}{0.92} = 1086.96$ W

Average speed of mover, $\quad u_a = 2l_s f_n = 2 \times 0.02 \times 60 = 2.4$ m/s

Average thrust, $\quad F_{xe} = \dfrac{P_e}{u_a} = \dfrac{1086.96}{2.4} = 452.9$ N

Now, for the entire surface area of the stator, the specific thrust f_x may be chosen to be between 0.5 and 2 N/cm². We choose f_x = 1.5 N/cm². Thus, referring to Fig. 8.9, we have

$$D_{es} = \frac{F_{xe}}{f_x(2\pi n l_s)} = \frac{452.9}{1.5 \times 2\pi \times 4 \times 0.02 \times 10^4} \approx 0.12 \text{ m}$$

For PMs, we choose Magnequench MQ3-F38, which has the following properties:

Specific weight, γ_{PM} = 7600 kg/m³

Residual flux density @25°C, B_r = 1.27 T

Coercive force @25°C, H_c = 11.7 kOe = 0.93153 MA/m

Recoil permeability, μ_{re} = 1.07 μ_0

Temperature coefficient of B_r (to 100°C), $\quad k_{B_r}$ = -0.09% per °C

Temperature coefficient of H_c (to 100°C), $\quad k_{H_c}$ = -0.6% per °C

Operating temperature, 75°C = $T°$.

Thus,

$$B_r @ 75°C = B_{ro} \left[1 + \frac{k_{B_r}}{100} (T° - 25°) \right]$$

$$= 1.27 \left[1 - \frac{0.09}{100} (75 - 25) \right] = 1.213 \text{ T} \qquad (8.28)$$

And

$$H_c @ 75°C = H_{co} \left[1 + \frac{k_{H_c}}{100} (T° - 25°) \right]$$

$$= 0.93153 \left[1 - \frac{0.6}{100} (75 - 25) \right] = 0.652 \text{ MA/m} \qquad (8.29)$$

The mechanical airgap is chosen from mechanical considerations. We may choose $g \approx (0.4 \text{ to } 1) \times 10^{-3}$ m for $P = 100$ to 1000 kW. For the present design we choose $g = 0.5 \times 10^{-3}$ m (or 0.5 mm).

Referring to Fig. 8.9, the PM radial thickness is obtained from $h_m \approx (4 \text{ to } 6)g$. Letting $h_m = 5g$, we obtain $h_m = 5 \times 0.5 \times 10^{-3} = 2.5 \times 10^{-3}$ m.

Next, the number of stator circumferential poles is given by $2p = 4, 6$. We let $2p = 4$.

Choose the pole span α_p (Fig. 8.9) from

$$\alpha_p = \frac{360}{2p} \times k_p \quad \text{with} \quad k_p = 0.5 \text{ to } 0.65 \qquad (8.30)$$

Assuming $k_p = 0.55$, for the present design we obtain

$$\alpha_p = \frac{360}{4} \times 0.55 = 50° $$

In order to determine the PM maximum and minimum flux, ϕ_{max} and ϕ_{min} respectively, per pole, the flux density distribution must be obtained. Approximately, the flux lines take the form shown in Fig. 8.13. Let the airgap permeances have a maximum and a minimum value, G_{max} and G_{min} respectively. These are calculated from

$$G_{max} = \mu_0 \pi \left[D_{is} - 2g + \frac{1}{2}(g + h_m) \right] \frac{l_s}{g + h_m} \times \frac{\alpha_p}{360}$$

$$= 4\pi \times 10^{-7} \pi \left[0.12 - 2 \times 0.0005 + \frac{1}{2}(0.0005 + 0.0025) \right] \frac{20}{(0.5 + 2.5)}$$

$$\times \frac{50}{360} = 0.44 \times 10^{-6} \text{ H} \qquad (8.31)$$

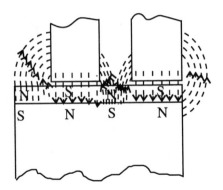

FIGURE 8.13
Approximate flux density distribution.

and

$$G_{min} = (1 + k_\sigma)(G_{min1} + G_{min2}) \tag{8.32}$$

where

$$G_{min1} = 3.3\,\mu_0 \times \frac{1}{2}(D_{is} - g + h_m)\frac{\alpha_p}{360}$$

$$= 3.3 \times 4\pi \times 10^{-7} \times \frac{1}{2}(0.12 - 0.0005 + 0.0025)\frac{50}{360}$$

$$= 0.0351 \times 10^{-6}\ \text{H} \tag{8.33}$$

Check if $l_s \le 2(g + h_m)$; if so, then $G_{min2} = 0$. Otherwise calculate G_{min2} from

$$G_{min2} = 4\mu_0 \left(\frac{1}{2}D_{is} - \sqrt{\frac{1}{2}l_s(g + h_m)}\right) \ln\left[\frac{l_s}{2(g + h_m)}\right]\frac{\alpha_p}{360}$$

$$= 4 \times 4\pi \times 10^{-7} \left(\frac{0.12}{2} - \sqrt{\frac{1}{2} \times 0.02\,(0.5 + 2.5)\,10^{-3}}\right)$$

$$\times \ln\left[\frac{0.02}{2(0.5 + 2.5)\,10^{-3}}\right]\frac{50}{360} = 0.04515 \times 10^{-6}\ \text{H} \tag{8.34}$$

Introducing the leakage coefficient k_σ to account for the extreme left and extreme right positions' flux lines, we write

$$G_{min} = (1 + k_\sigma)(G_{min1} + G_{min2}) \qquad (8.35)$$

Assuming $k_\sigma = 0.5$, we obtain

$$G_{min} = (1 + 0.5)(0.0351 + 0.04515) 10^{-6}$$

$$= 0.12 \times 10^{-6} \text{ H}$$

Now, the PM (per pole) may be replaced by a fictitious mmf I_{PM} such that

$$I_{PM} = \frac{B_r h_m}{\mu_{re}} = \frac{1.213 \times 2.5 \times 10^{-3}}{1.07 \times 4\pi \times 10^{-7}} = 2255 \text{ A}$$

For the given large airgap, the saturation factor is expected to be low. So we choose $k_s = 0.025$. Consequently, the maximum and minimum values of the flux are given by

$$\phi_{max} = \frac{G_{max} I_{PM}}{1 + k_s} = \frac{0.44 \times 10^{-6} \times 2255}{1 + 0.025}$$

$$= 0.968 \times 10^{-3} \text{ Wb/pole}$$

and

$$\phi_{min} = \frac{G_{min} I_{PM}}{1 + k_s} = \frac{0.12 \times 2255 \times 10^{-6}}{1 + 0.025}$$

$$= 0.264 \times 10^{-3} \text{ Wb/pole}$$

Thus, we obtain the maximum flux through a pole, $(\phi_{max})_p$, from

$$(\phi_{max})_p = \phi_{max} - \phi_{min} = (0.968 - 0.264) 10^{-3}$$

$$= 0.704 \times 10^{-3} \text{ Wb}$$

The variations of fluxes in various regions of interest are as follows. In the coil, the flux varies from $+n(\phi_{max})_p$ to $-n(\phi_{max})_p$. The flux in a pole varies from $(\phi_{max} - \phi_{min})$ to ϕ_{min}. Finally, the flux in the mover core varies from $(\phi_{max} - \phi_{min})$ to $-(\phi_{max} - \phi_{min})$. The PM induced voltage E_1 is given by the emf equation

$$E_1 = 2pn(\phi_{max} - \phi_{min}) \frac{\omega_1}{\sqrt{2}} k_{re} N_c = C_e N_c \qquad (8.36)$$

where $k_{re} \approx 0.9$ to 0.95, which is a reduction factor owing to waveform distortion. We choose $k_{re} = 0.95$ and C_e is a constant given by

$$C_e = 2pn(\phi_{max} - \phi_{min}) \frac{\omega_1}{\sqrt{2}} k_{re}$$

$$= 2 \times 2 \times 2 \, (0.968 - 0.264) \, 10^{-3} \times \frac{377}{\sqrt{2}} \times 0.95$$

$$= 1.43 \text{ V} \tag{8.37}$$

The mmf or ampere-turns per pole is

$$N_c I_n = \frac{P_e}{C_e} = \frac{1086.96}{1.43} = 760 \text{ At/pole} \tag{8.38}$$

We choose a rated current density from $J_{co} \approx (2 \text{ to } 4.5)10^6$ A/m^2 and a window fill factor from $k_{fill} \approx 0.55$ to 0.7. Let $J_{co} = 3.13 \times 10^6$ A/mm^2 and $k_{fill} = 0.65$. The active (copper) area of a coil can be expressed as (Fig. 8.9)

$$A_{co} = a \times h_{coil} = \frac{N_c I_n}{J_{co}} \tag{8.39}$$

Substituting numerical values yields

$$A_{co} = \frac{760}{3.13} \times 10^{-6} \approx 243 \times 10^{-6} \text{ m}^2 = 243 \text{ mm}^2$$

The required window area becomes

$$A_w = \frac{A_{co}}{k_{fill}} = \frac{243}{0.65} = 374 \text{ mm}^2$$

The available coil width a (Fig. 8.9) in the window area is given by

$$a \approx \frac{1}{2} D_{is} \left(\sin \frac{90°}{2} - \sin \frac{\alpha_p}{2} \right)$$

$$= \frac{0.12}{2} \left(\sin 45° - \sin \frac{50°}{2} \right) = 17.06 \times 10^{-3} \text{ m}$$

Referring to Fig. 8.9, the coil height becomes

$$h_{coil} = \frac{A_w}{a} = \frac{374 \times 10^{-6}}{17.06 \times 10^{-3}} = 21.92 \text{ mm} \tag{8.40}$$

Now, the outer diameter D_{PO} (Fig. 8.9) is given by

$$D_{PO} = D_{is} + 2h_{coil} + 0.003$$

$$= 0.12 \times 2 \times 21.92 \times 10^{-3} + 0.003 = 166.84 \times 10^{-3} \text{ m} \tag{8.41}$$

The pole width b_p (Fig. 8.9) is

$$b_p = D_{is} \sin \frac{\alpha_p}{2} = 0.12 \sin \frac{50°}{2} = 50.7 \times 10^{-3} \text{ m}$$

In order to determine the core depth (or the thickness of the yoke) a core flux density in the range $B_{core} \simeq 1.1$ to 1.35 T may be assumed. We choose $B_{core} = 1.3$ T. Consequently,

$$h_{core} \simeq \frac{\phi_{max}}{2B_{core}l_s} = \frac{0.968 \times 10^{-3}}{2 \times 1.3 \times 0.02} = 18.62 \times 10^{-3} \text{ m} \tag{8.42}$$

Therefore, the external diameter of the stator D_{es} (Fig. 8.9) becomes

$$D_{es} = D_{PO} + 2h_m = (166.84 + 2 \times 18.62)10^{-3} \text{ m} = 204.1 \text{ mm}$$

The elements of the equivalent circuit of Fig. 8.12 are now determined. The two components of the inductance are expressed as

$$L_m = C_m N_c^2 \quad \text{and} \quad L_\sigma = C_\sigma N_c^2 \tag{8.43}$$

where

$$C_m = n(G_{max} + G_{min}) 2p$$

$$= 2(0.44 + 0.12) 10^{-6} \times 2 \times 2$$

$$= 4.48 \times 10^{-6} \text{ H} \tag{8.44}$$

and

$$C_\sigma = 4p(\lambda_s l_{cs} + \lambda_e l_{ec})\mu_0 \tag{8.45}$$

where l_{cs} = coil length in a slot

$$= 2nl_s = 2 \times 2 \times 0.02 = 0.08 \text{ m}$$

l_{es} = length of the end connection

$$= b_p + \frac{\pi a}{2} = 0.0507 + \frac{\pi \times 0.017}{2} = 0.0774 \text{ m}$$

λ_s = slot permeance

$$= \frac{h_{coil}}{3a} = \frac{21.92}{3 \times 17.06} = 0.4283 \text{ H}$$

λ_e = end-connection permeance

$$= \frac{1}{2}\lambda_s = 0.2146 \text{ H}$$

Hence, with the above numerical values we obtain

$$C_\sigma = 4 \times 2(0.4283 \times 0.08 + 0.2146 \times 0.0774)4\pi \times 10^{-7}$$

$$= 0.511 \times 10^{-6} \text{ H}$$

Defining $C_L = C_m + C_\sigma$, with $L_s = C_L N_c^2$ we obtain

$$C_L = (4.48 + 0.511)10^{-6} = 4.99 \times 10^{-6} \text{ H}$$

As denoted previously, in the above equations N_c is the number of turns per coil. With all the coils connected in series, the stator resistance is given by (8.22), or $R_s = C_R N_c^2$. With $N_c I_n = 760$ and $l_c = 2(0.08 + 0.0774) = 0.3148$ m, we obtain

$$C_R = \frac{4 \times 0.3148 \times 2.1 \times 10^{-8} \times 3.13 \times 10^{6}}{760}$$

$$= 1.089 \times 10^{-4} \ \Omega$$

The copper loss P_{co} is obtained from

$$P_{co} = I_n^2 R_s = C_R(N_c I_n)^2 = 1.089 \times 10^{-4}(760)^2$$

$$= 62.9 \text{ W}$$

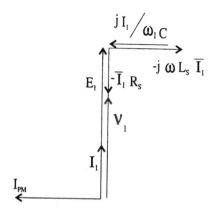

FIGURE 8.14
Phasor diagram.

We assume core losses P_{core} to be about 10% of the copper loss. Thus,

$$P_{core} = 0.1 \times 62.9 = 6.29 \text{ W} \approx 6.3 \text{ W}$$

Finally, mechanical (and other) losses are given by

$$P_{mech} = P_e - P_{0e} - P_{co} - P_{core}$$

$$= 1086 - 1000 - 62.9 - 6.29 = 16.81 \text{ W}$$

As this value is greater than 2% of the rated output, we consider it practical and the desired efficiency is achievable.

With a tuning capacitor $C = 1/\omega_1 L_s = 178\mu F$, the phasor diagram for the LA on a resistive load becomes as shown in Fig. 8.14, from which

$$\bar{E}_1 = \bar{V}_1 + R_s \bar{I}_n \tag{8.46}$$

Consequently, the number of turns per coil is given by

$$N_c = \frac{V_n}{C_e - C_R(N_c I_n)}$$

$$= \frac{120}{1.43 - 1.089 \times 10^{-4} \times 760} = 89 \text{ turns} \tag{8.47}$$

With $N_c = 89$, $I_n = 760/89 = 8.54$ A. Assuming an allowable current density $J_c = 3.13 \times 10^6$ A/m^2 the conductor diameter is

$$d_c = \sqrt{\frac{4I_n}{\pi J_c}} = \sqrt{\left(\frac{4 \times 8.54}{\pi \times 3.13 \times 10^6}\right)} = 1.86\text{mm}$$

So wire size AWG-12 may be chosen to wind the coils.

8.4 STABILITY CONSIDERATIONS

For many applications, the free-piston Stirling engine has been chosen as the prime mover to drive the LA. The Stirling engine is used as an energy conversion device converting thermal energy to mechanical motion. Under proper conditions, sustained oscillations of the pistons of the engine occur, and electric power can be derived from an LA whose moving part is mechanically connected to the power piston of the engine.

The engine consists of two moving parts, namely a displacer piston and a power piston. The two pistons interact with each other dynamically through the engine working gas. The motion of the power piston causes changes in working gas pressure, which produces motion of the displacer. The displacer transfers the working gas between hot and cold spaces, thus changing working gas pressure and hence the force on the piston. Under certain conditions, sustained oscillations of the piston will occur. Electric power can be generated at the output winding of the LA through magnetic induction principles. As the operating principle of the system is based on mechanical oscillations, the stability of the system becomes crucial for proper operation. The suitable stability criterion is that the rate of change in power generation by the piston be slower than the rate of change in power dissipation and power output, such that no excess power is accumulated in the system and the amplitude of oscillations will be bounded within the mechanical stroke limit.

The condition for sustained oscillation, which is a design criterion of the engine, provides the lower limit for stable operation, since the engine would not run otherwise. The upper limit of stable operation can be determined from the stability criterion stated above, which depends on the power dissipation of the system, output power levels, and oscillation frequency.

The stability criterion may also be stated from an alternative viewpoint. During the operation of the system, the rate of change of engine power generated is primarily a function of the rate of change in the piston stroke and is defined by the piston time constants (piston mass and stiffness). The rate of change of power dissipation is a function of the rate of change of the load current and is defined by the electrical time constant (inductance and resistance of the alternator). Therefore, for a stable operation, the piston time constant should be larger than the electrical time constant of the alternator. A means to increase the piston time constant is to increase its mass, although this is not practical because it will result in higher gas spring losses and larger gas spring piston area and volume. Therefore, the practical and economical approach is to minimize the electrical time constant. Decreasing the electrical time constant of the alternator requires lowering the alternator inductance, which means that magnet volume should be increased. Increasing the magnet volume might not be desirable due to the high cost of the magnet material. An alternative approach to decreasing the electrical time constant is to add a capacitor in the output winding. Of course, this additional capacitor also serves other functions. Considering the costs associated with these two solutions, an optimal ratio of the magnet volume and capacitor size may be expected.

The schematic model of the free-piston Stirling engine and LA system under study is shown in Fig. 8.15. The dynamic equation of the engine has been studied by Redlich and Berchowitz [3]. However, in their study, the dynamics of the alternator was not considered. For the displacer piston, the dynamic equation can be written as

$$M_d \ddot{X} + D_d \dot{X}_d = A_d(P_d - P) \tag{8.48}$$

and that for power piston as

$$M_p \ddot{X}_p + D_p \dot{X}_p + F_c + K_p X_p + (A - A_d) \frac{\partial P}{\partial X_d} X_d = 0 \tag{8.49}$$

where, A_d = displacer rod area, m²
D_d = displacer damping constant, N/m × s
P_d = gas spring pressure, N/m²
P = working gas pressure, N/m²
X_d = displacer displacement, m
X_p = power piston displacement, m
A = cylinder area
F_c = electromagnetic force from the alternator, N
M_d = displacer mass, kg
M_p = power piston mass, kg

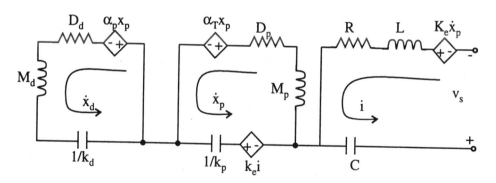

FIGURE 8.15
Dynamic model of engine/alternator system.

In order to write the dynamic equation in terms of the displacements X_d and X_p, (8.48) and (8.49) are linearized to obtain

$$M_d \ddot{X}_d + D_d \dot{X}_d = A_d \left[\left(\frac{\partial P_d}{\partial X_d} - \frac{\partial P}{\partial X_d} \right) X_d - \frac{\partial P}{\partial X_p} X_F \right] = -K_d X_d - \alpha_p X_p \quad (8.50)$$

and

$$M_p \ddot{X}_p + D_p \dot{X}_p + F_c + K_p X_p + \alpha_T X_d = 0 \quad (8.51)$$

where

$$K_d = -A_d \frac{\partial P_d}{\partial X_d} - \frac{\partial P}{\partial X_d}$$

$$\alpha_p = \frac{\partial P}{\partial X_p} A_d$$

$$K_p = (A - A_d) \frac{\partial P}{\partial X_p}$$

$$\alpha_T = (A - A_d) \frac{\partial P}{\partial X_d} X_d$$

The coefficient M's, D's, K's and α's are determined from design data.

The structure of the PMLA used in the system is shown in Fig. 8.4. The output winding is inserted in the slots of outer stator and inner stator. The magnets, one center magnet and two spring magnets, are assembled together to form the plunger. The electrical equation for the alternator, including the tuning capacitor, is

$$Ri + L \frac{di}{dt} + \frac{1}{C} \int i \, dt = e_g - V_t \quad (8.52)$$

where, R = winding resistance, Ω
L = winding inductance, H
i = instantaneous output current, A
C = tuning capacitance, F
e_g = instantaneous induced voltage, V
V_t = winding terminal voltage, which may be the grid voltage if connected to the grid, or load voltage if connected to a load

The electromechanical energy conversion equation is

$$F_e X_P = e_g i \tag{8.53}$$

To determine electromagnetic force F_c and induced voltage e_g, we first express the flux linkage of the stator winding $\phi(x)$ as

$$\phi(x) = \pi D_m B_m (N_t + N_c) X_p \tag{8.54}$$

where B_m = magnetic flux density in the airgap due to magnet only, T
D_m = mean diameter of the magnet, m
N_i, N_o = number of turns for inner and outer stator winding, respectively

The no-load induced voltage e_g is therefore

$$e_g = \frac{-d\phi}{dt} = -\frac{d\phi}{dX_p} \dot{X}_p = -\pi D_m B_m (N_i + N_o) \dot{X}_p \tag{8.55}$$

From (8.54) and (8.55), if the displacement X_p varies sinusoidally in time, the flux linkage and induced voltage also vary sinusoidally.

The electromagnetic force developed F_e is

$$F_e = i \frac{\partial\phi}{\partial X_p} = i\pi D_m (N_i + N_o) B_m \tag{8.56}$$

Let

$$k_m = \pi D_m (N_i + N_o) B_m \tag{8.57}$$

Then (8.55) and (8.56) can be written as

$$e_g = -k_m \dot{X}_p \tag{8.58}$$

$$F_e = k_m i \tag{8.59}$$

From (8.55), the rms value of induced voltage is

$$E_{rms} = \sqrt{2}\, \pi D_m B_m (N_i + N_o)\pi f_1 L_{stroke}$$

$$= \sqrt{2}\, k_m \pi f_1 L_{stroke} \tag{8.60}$$

where f_i is the oscillation frequency of the plunger and L_{stroke} is the stroke length of the plunger.

Based on the dynamic equations presented above, the dynamic model of the engine/alternator system is developed as shown in Fig. 8.15, where the diamond voltage source denotes a controlled voltage source.

For sustained oscillations, the frequencies of oscillations of the displacer piston, the power piston, and the electric current as well as the induced voltage are the same, say ω. If the system is connected to the grid, ω will be the same as the grid frequency. Under steady state, we may assume the solutions to the dynamic equations as follows:

$$v_t = \sqrt{2} \, V_t \cos \omega t$$

$$e_g(t) = \sqrt{2} \, E_{rms} \cos(\omega t + \delta)$$

$$i(t) = \sqrt{2} \, I \cos(\omega t + \phi)$$

$$X_d(t) = X_d \cos(\omega t + \phi_d)$$

$$X_p(t) = X_p \cos(\omega t + \phi_p)$$

where ϕ is the power factor angle and δ the power angle. Then, the dynamic equations become

$$(-M_d\omega^2 + j\omega D_d + k_d)X_d + \alpha_p X_p = 0 \tag{8.61}$$

$$(-M_p\omega^2 + j\omega D_p + k_p)X_p + \alpha_T X_d + F_e = 0 \tag{8.62}$$

$$RI + j\left(\omega L - \frac{1}{\omega C}\right)I = E_g - V_t \tag{8.63}$$

Solving (8.61) and (8.62) we obtain

$$\frac{X_d}{X_p} = \frac{-\alpha_p}{k_d - M_d\omega^2 + j\omega D_d} = \frac{|X_d|}{|X_p|} \angle \theta \tag{8.64}$$

The magnitudes of stroke of the displacer and power piston are, respectively,

$$|X_d| = \frac{\alpha_p k_m V_t}{2\sqrt{R^2 + \left(\omega L - \frac{1}{\omega C}\right)^2} \, |det|} \tag{8.65}$$

$$|X_p| = \frac{k_m V_t \sqrt{(k_d - M_d\omega^2)^2 + (\omega D_d)^2}}{2\sqrt{R^2 + \left(\omega L - \dfrac{1}{\omega C}\right)^2} \, |det|} \tag{8.66}$$

where

$$|det| = \begin{vmatrix} k_d - M_d\,\omega^2 + j\omega D_d & \alpha T \\[2mm] & k_p - M_p\omega^2 + j\omega D_p \\[2mm] \alpha_T & \dfrac{jk_m^2\omega/4}{R + j\left(\omega L - \dfrac{1}{\omega C}\right)} \end{vmatrix}$$

The average power developed by the engine, P_d, is

$$P_d = \langle(k_p X_p - \alpha_T X_d)\dot{X}_p\rangle \tag{8.67}$$

Since X_d and X_p are sinusoidal in time under steady state, (8.67) becomes

$$P_d = -j\omega \frac{|\alpha_T|}{4} (X_d^* X_p - X_d X_p^*) \tag{8.68}$$

or

$$P_d = \frac{\omega \, |\alpha_T|}{2} |X_d| \, |X_p| \sin\theta \tag{8.69}$$

The average rate of power generated over one cycle of ω is

$$W_d = P_d \frac{2\pi}{\omega} = \pi |\alpha_T| |X_d| |X_p| \sin\theta \tag{8.70}$$

The power removal from the shaft is the sum of the dissipated power and the output power of the alternator, and is expressed as

$$P_c = P_{diss} + P_{out}$$

$$= (P_{diss})_{displacer} + (P_{diss})_{piston} + (P_{diss})_{alternator} + P_{out} \tag{8.71}$$

where $(P_{diss})_{displacer}$ = power dissipated by the displacer

$$= D_d(\omega |X_d|)^2$$

$(P_{diss})_{piston}$ = power dissipated by the piston

$$= D_p(\omega |X_p|)^2$$

$(P_{diss})_{alternator}$ = core losses and copper losses of the alternator

P_{out} = output power of the alternator

The estimated core and copper losses in the alternator are about 6% of the output power.

The average rate of total power removal from the shaft over one cycle of ω is

$$W_c = P_c \frac{2\pi}{\omega} \tag{8.72}$$

Therefore, mathematically, the stability criterion can be stated as

$$W_d \leq W_c \tag{8.73}$$

Substituting (8.69) through (8.72) into (8.73), for stable operation of the system, we must have

$$20 \log \left| \frac{\frac{\omega |\alpha_T|}{2} |X_d| |X_p| \sin \theta}{D_d(\omega |X_d|)^2 + D_p(\omega |X_p|)^2 + 1.06 P_{out}} \right| \leq 0 \tag{8.74}$$

Expression (8.74) is the stability criterion in terms of system parameters and dynamic variables, and provides the upper limit for stability of the system.

The stability criterion developed in previous sections is now applied to a practical free-piston Stirling engine and PMLA power generation system. The calculated and design parameters pertaining to such a system are given in Table 8.1.

In the study of the frequency range of stable operation, the fact that the oscillating frequency of the system cannot be arbitrarily imposed externally once the system is built needs to be particularly considered. After the system is designed, the natural oscillating frequencies of the piston, $\omega_d (= \sqrt{k_d/M_d})$ and $\omega_p (= \sqrt{k_p/M_p})$, are fixed. The oscillating frequency of the system is determined by [3]

TABLE 8.1 Parameters and Data of the Power Generation System

Components	Parameters	Calculated Values
Alternator	R	0.04239 Ω
	L	9.422 mH
	C	512 μF
	$N_i + N_o$	52 turns
	B_m	0.875 Tesla
	D_m	0.366 m
Piston	M_d	6.45 kg
	M_p	23.0 kg
	D_d	87.3 Ns/m
	D_p	93.45 Ns/m
	k_d	770.3 × 10³
	k_p	3885.2 × 10³
	α_p	1496.7 × 10³
	α_T	−1943.85 × 10³

$$\omega = \frac{\omega_d \omega_p (Q_p + Q_d)}{\omega_p Q_d + \omega_d Q_p} \tag{8.75}$$

$$Q_d = \frac{\omega_d}{2\pi} \frac{M_d}{D_d}$$

where

$$Q_p = \frac{\omega_p}{2\pi} \frac{M_p}{D_p}$$

Therefore, from (8.75), the system oscillation frequency ω can be varied by changing the output power of the alternator, which is equivalent to changing Q_p. For a normal design, since oscillating frequency ω lies between ω_d and ω_p, ω_d is not equal to ω_p. However, the engine system is designed such that ω_d is close to ω_p to eliminate the loading effect of the output power level on oscillating frequency. This indicates that, once the system is designed, ω does not change much even when the load varies.

As the variation of load (output level of the alternator) can provide only a very narrow oscillating frequency band, we choose to vary ω_d to change the oscillating frequency ω in order to investigate a relatively broader frequency band. The results are plotted in Fig. 8.16 in terms of the magnitude of 20 log(P_d/P_c) versus frequency for different levels of output power. When 20 log(P_d/P_c) exceeds zero, the system begins to enter the unstable operating frequency region. Figure 8.16 shows that the system is stable at low frequency (≤55 Hz) and at higher frequency (≥95 Hz). However, when the frequency is below $\omega_d/2\pi$ (≈55 Hz), oscillations cannot be sustained. Therefore,

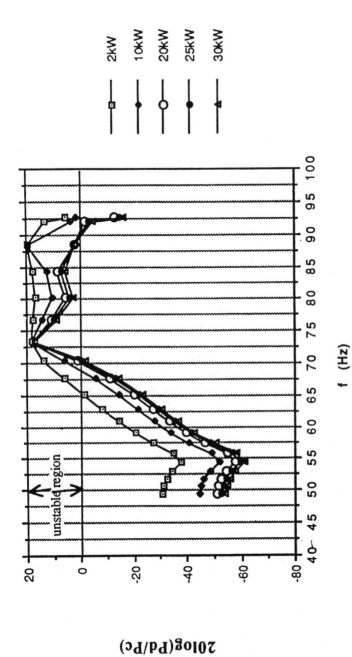

FIGURE 8.16
Bode plot for stability analysis.

TABLE 8.2 Stability Limits.

Stable Frequency Range (Hz)	Output Power (kW)
55 to 65	2
55 to 68	10
55 to 71	20
55 to 71	25
55 to 71	30

$\omega_d/2\pi$ is considered the lower limit of stable frequency region. Similarly, if the frequency is too high (≥ 95 Hz), the engine will not run. Based on Fig. 8.16 and the above discussion, the stability limit of the system in terms of operating frequency is presented in Table 8.2 [4].

8.5 CHOICE OF TUNING CAPACITOR

A practical concern relating to an LA engine system such as that discussed in the preceding section is to minimize the total cost of the magnets and the tuning capacitor, as both the PM material used in the LA and the high power rating capacitor for tuning are expensive. Therefore, from an economic standpoint, the cost trade-off of magnet and tuning capacitor is important in the design of the PMLA.

In considering the cost trade-off between magnet and capacitor, criteria in selecting capacitor size must ensure a proper operation of the power generation system. Then, the cost estimates for each pair of magnet material and capacitor can be made and compared. Cost minimum is obtained from the table of comparison as a result of the above study or by an optimization procedure.

The turning capacitor is used for the following purposes:

1. *For the stability of the engine/alternator system:* The stability of the system requires that the rate of power removal from the engine be faster than the rate of power generated by the piston. This suggests that the output power of the alternator should be maximized through the choice of the capacitor to ensure stable operation.

2. *For prevention of magnet demagnetization:* When the mmf generated by the armature reaction is greater than the PM operating magnetic strength ($H_m h_m$), the PM will be demagnetized. This suggests that the armature reaction field should be limited by the choice of a proper capacitor under normal operating conditions.

3. *For power factor correction:* Because of the lagging power factor load, ($\omega L - 1/\omega C$) should be negative to provide a leading power factor angle in order to improve the overall power factor of the alternator.

These three requirements are the criteria for sizing the capacitor. Though power factor correction is essential to enhance the performance of the alternator, items 1 and 2 are crucial for proper operation of the system. In other words, if the size of the capacitor cannot meet the requirements of 1 and 2, there exists a danger of loss of PM magnetic strength, or the engine will be destroyed due to unbounded amplitude of piston oscillations. Therefore, the size of the tuning capacitor should be selected to meet the requirements of 1 and 2 first; then within the permissible range determined by items 1 and 2, we choose the capacitor to provide a better overall power factor.

In the trade-off of magnet cost versus capacitor cost, the emphasis should be on satisfying the criteria for preventing PM demagnetization and securing stability. To be precise, changing the PM volume will inevitably result in a change in the alternator inductance in order to maintain the level of output power. The corresponding range of capacitor size required should be determined on the basis of 1 and 2. Then, within that determined range of capacitor size, the one which provides a better power factor is selected. See Reference [5] for details.

REFERENCES

1. I. Boldea and S. A. Nasar, "Permanent magnet linear alternator: Part I. Fundamental equations," *IEEE Trans.*, vol. AES-23, 1987, pp. 73-78.
2. I. Boldea and S. A. Nasar, "Permanent magnet linear alternator: Part II. Basic design guidelines," *IEEE Trans.*, vol. AES-23, 1987, pp. 79-82.
3. R. W. Redlich and D. W. Berchowitz, "Linear dynamics of free-piston Stirling engine," *Proc. I Mech. E.*, March 1985, pp. 203-213.
4. Z. X. Fu, S. A. Nasar, and M. Rosswurm, "Stability analysis of free-piston Stirling engine power generation system," 27th Intersociety Energy Conversion Engineering Conference, San Diego, CA, August 3-7, 1992.
5. Z. X. Fu, S. A. Nasar, and M. Rosswurm, "Tradeoff between magnet and tuning capacitor in free-piston Stirling engine power generation system," 27th Intersociety Energy Conversion Engineering Conference, San Diego, CA, August 3-7, 1992.

INDEX

Ampere's law, 5, 6, 16, 38

B-H curve, 8, 9,18, 23, 25

Carter's coefficient, 46
Continuity equation, 6
Core loss, 11
Coupling coefficient, 19
Curie temperature, 11

Demagnetization curve, 11, 23, 116
Displacement current, 6

Eddy-current loss, 11, 12
Electric field, 2
Electrical force, 26-30
Electromagnetic field theory, 2-7
Emf, 3
    motional, 5
    transformer, 7
Energy conservation, 28
Energy product, 11, 23

Faraday's law, 3, 4
Force calculations, 26-30
Force equation, 27
Fringing, 15, 18

Gauss' law, 6

Hysteresis loop, 10, 11
Hysteresis loss, 11, 12

Inductance, 19

Laminated core, 12
LEA, 33
    application, 40
    classification, 36, 40
    operation, 36
    permanent magnet, 40
    reluctance, 40
    solenoid, 40
    stepper, 40
    switched reluctance, 41

Leakage factor, 24
Leakage flux, 14, 15, 18
Lenz's law, 3
Linear electric actuator, *see* LEA
LEG, 33, 36
    classification, 36, 40
    configurations, 36, 40
    design, 214-223
    fields, 227
    model, 224
    moving coil, 202
    moving iron, 203, 211
    moving magnet, 202
    operation, 39, 201
    permanent magnet
      (PMLA), 202
    permeance function, 213
    stability, 223
    tuning capacitor, 232
Linear electric generator *see* LEG
Linear induction actuator (LIA), 45
    attraction force, 60, 61
    construction, 46
    design, 75-88
    equivalent circuits, 58, 59, 63,
      76
    field distributions, 54-58
    flat, 47-49
    performance, 45, 61, 65, 68
    power flow, 59
    state-space equations, 68
    thrust, 60, 61
    tubular, 47, 50, 51
    vector control, 70
    windings, 51-54
Linear permanent magnet synchronous
      actuators (LPMSA), 91
    air core, 102
    construction, 91
    control, 105-112
    design, 113-134
    dynamics, 105
    fields in, 92, 94
    generator, 99
    losses, 122
    model, 94, 97, 102
    motor, 99
    normal force, 101, 104, 105
    parameters, 125

    rectangular current excitation,
      102, 107, 126
    sinusoidal current excitation,
      94, 109
    thrust, 101, 105, 120
Linear reluctance synchronous
      actuators (LRSA), 135
    configurations, 136
    control, 147-152
    design, 152-161
    equivalent circuits, 148
    fields, 141
    flat, 140
    magnetizing inductance, 138
    model, 144
    normal force, 146, 147
    performance, 147-160
    pole pitch, 144
    saliency coefficients, 136, 142,
      143
    thrust, 145, 147
    transversely laminated
      secondary, 139, 140
    tubular, 140
Linear stepper actuators (LSAs), 179
    advantages, 179
    airgap permeance, 185, 186
    configurations, 180
    control, 188-193
    design, 193-199
    disadvantages, 179
    equivalent circuit, 187
    flat, 180, 182
    hybrid, 180, 183
    model, 184
    normal force, 184
    static forces, 183
    thrust, 181, 184
    variable reluctance, 180
Linear switched reluctance actuators
      (LSRA), 163
    configurations, 163
    control, 172
    design, 173-178
    flat, 164, 165
    normal force, 166
    state equations, 172
    thrust, 167
    tubular, 165, 166

Lorentz force, 2, 5, 7, 37

Magnetic circuits, 13-18
Magnetic coenergy, 30
Magnetic field, 2, 5
Magnetic flux, 3, 14
Magnetic flux density, 2
Magnetic losses, 11
Magnetic materials, 7-11, 17
    diamagnetic, 8
    nonmagnetic, 8
    paramagnetic, 8
Magnetic potential, 13
Magnetic saturation, 8, 18
Magnetic stored energy, 19, 20
Magnetomotive force, 13
mmf, 13, 16
Maxwell's equation, 2, 6

Ohms' law, 7

Permanent magnet, 22-26
Permeability, 7
    of free space, 7, 8
    incremental, 9
    nonlinear, 8, 9
    recoil, 24
    relative, 9, 10
Permeance, 14, 16, 20, 23
Permeance coefficient, 23
Permittivity, 7
Potential difference, 2, 3

Recoil permeability, 26
Reluctance, 14, 16

Sawyer motor, 39
Slot inductance, 20-22
Stacking factor, 12, 17, 50
Stokes' theorem, 4